T0305855

Injection Molding Process Modelling

Injection Molding Process Modelling presents the application of CAE, statistics and AI in defect identification, control, and optimization of injection molding process for quality production. It showcases CAE in determining the optimal placement of injection points, designing cooling channels, and ensuring that the mold will produce parts with the desired specifications. The book illustrates the capability of the CAE tools to simulate molten plastic flow within a mold during the injection molding process.

Explaining how the use of CAE, statistical tools and AI enhances efficiency, accuracy, and collaboration, the book explores the contributions to injection molding in product design and visualization; prototyping and testing; mold design; and analysis and simulation. It emphasizes the integration of statistical tools for optimized efficiency and waste reduction, including statistical process control (SPC), Design of Experiments (DOE), Regression Analysis, Capability Indices, Interaction effects, and many more. The book also illustrates the predictive modelling of typical injection molded product defects using intelligent algorithms.

The book will interest industry professionals and engineers working in manufacturing, production, automation, and quality control.

Injection Molding Process Modelling

Statistics, CAE, and AI Applications

Tien-Chien Jen
Edwell Tafara Mharakurwa
Steven Otieno Otieno
Fredrick Madaraka Mwema
Job Maveke Wambua

CRC Press
Taylor & Francis Group
Boca Raton New York London

CRC Press is an imprint of the
Taylor & Francis Group, an **informa** business

First edition published 2025
by CRC Press
2385 NW Executive Center Drive, Suite 320, Boca Raton FL 33431

and by CRC Press
4 Park Square, Milton Park, Abingdon, Oxon, OX14 4RN

CRC Press is an imprint of Taylor & Francis Group, LLC

© 2025 Tien-Chien Jen, Edwell Tafara Mharakurwa, Steven Otieno Otieno, Fredrick Madaraka Mwema, and Job Maveke Wambua

ISBN: 978-1-032-79520-1 (hbk)
ISBN: 978-1-032-79522-5 (pbk)
ISBN: 978-1-003-49249-8 (ebk)

DOI: 10.1201/9781003492498

Typeset in Palatino
by SPi Technologies India Pvt Ltd (Straive)

"To the visionary pioneers of artificial intelligence; your contributions accelerate the achievement of a more efficient, sustainable, and interconnected future."

TC Jen, ET Mharakurwa, SO Otieno, FM Mwema, JM Wambua

"To my beloved late mother Eunice. M; your zeal for knowledge gaining continues to live within me, continue to rest in peace mum."

ET Mharakurwa

"To my younger siblings, may this work remind you that the greatest adventures lie within your dreams"

SO Otieno

"To my mother, *Mwende Mwema*; your life simplicity is an inspiration to me - live longer *mami*"

FM Mwema

"To my son, *Wambua Maveke*; may this work be an inspiration to always aim high and be at your best"

JM Wambua

Contents

About the Authors

Tien-Chien Jen

Prof. Jen joined the University of Johannesburg in August 2015, before he was a faculty member at the University of Wisconsin, Milwaukee. Prof Jen received his PhD in Mechanical and Aerospace Engineering from UCLA, specializing in thermal aspects of grinding. He has received several competitive grants for his research, including those from the US National Science Foundation, the US Department of Energy, and the EPA. Prof. Jen has brought in $3.0 million of funding for his research and has received various awards for his research including the NSF GOALI Award. Prof. Jen has established a Joint Research Centre with the Nanjing Tech University of China on the "Sustainable Materials and Manufacturing." Prof. Jen is also the Director of the newly established Atomic Layer Deposition Research Centre of the University of Johannesburg. SA National Research Foundation has awarded Prof. Jen an NNEP grant (National Nano Equipment Program) worth USD 1 million to acquire two state-of-the-art Atomic Layer Deposition (ALD) Tools for ultra-thin film coating. These two ALD tools will be the first in South Africa and possibly the first in the African continent. In 2011, Prof. Jen was elected as a Fellow of the American Society of Mechanical Engineers (ASME), which recognized his contributions to the field of thermal science and manufacturing. As stated in the announcement of Prof. Jen Fellow's status at the 2011 International Mechanical Engineering and Congress Exposition, it states "Tien-Chien Jen has made extensive contributions to the field of mechanical engineering, specifically in machining processes. Examples include, but are not limited to, environmentally benign machining, atomic layer deposition, cold gas dynamics spraying, fuel cells and hydrogen technology, batteries, and material processing." In addition, he was inducted into the Academy of Science of South Africa (ASSAf) in 2021. The Academy of Science of South Africa (ASSAf) aspires to be the apex organization for science and scholarship in South Africa, recognized and connected both nationally and internationally. Prof. Jen has written over 360 peer-reviewed articles, including 180 peer-reviewed journal papers, published in Elsevier, ACS, Taylor and Francis, Springer Nature, Wiley, and MDPI etc., some of the journals include Nano-Micro Letters, Coordination Chemistry Reviews, Nanotechnology Review, Journal of Molecular Liquids, Journal of Materials Research and Technology, Scientific Reports, International Journal of Heat and Mass Transfer, ASME Journal of Heat Transfer, ASME Journal of Mechanical Design and ASME Journal of Manufacturing Science and Engineering etc. He has written 16 book chapters and has 5 books published with the latest book titled "Chalcogenide: Carbon Nanotubes and Graphene Composites," published by CRC press-Taylor Francis, January 2021.

Edwell Tafara Mharakurwa

Dr Mharakurwa is a lecturer in the Department of Electrical and Electronic Engineering at Dedan Kimathi University of Technology (DeKUT). He also served as a teaching assistant (2011-2012) and lecturer (2015-2016) in the Department of Mechatronics Engineering at Chinhoyi University of Technology, Zimbabwe. Currently, he is the Chair of the Electrical and Electronic Engineering department, a position he held since 2020 at DeKUT. He also serves as an external examiner (since 2021) and has also assessed several postgraduate theses and dissertations. He has also voluntarily reviewed several journal and conference articles for different journal publishers. Dr Mharakurwa has supervised more than 80 undergraduate students and 5 MSc engineering students to completion. He is currently supervising 1 PhD student and 6 MSc engineering students. Dr Mharakurwa has published several peer-reviewed journals and conferences in his area of research interest not limited to the application of artificial intelligence in process control, power systems, electrical machines faults diagnosis and condition monitoring, residual life estimation of electrical machines etc.

Steven Otieno Otieno

Mr Otieno is currently a teaching assistant (since November 2023) and has been a research assistant in the Department of Mechanical Engineering at Dedan Kimathi University of Technology for three years since 2021. Having successfully completed a masters of science degree course in machine tools design and manufacturing engineering, his research area has been on advanced manufacturing, specifically interested in plastic processing through injection moulding. He has an ardent interest in intelligent predictive modelling and control of plastic injection moulding defects. He has presented his research in two peer-reviewed conferences and has published three articles in peer reviewed journals. His outstanding work on Computer Aided Engineering (CAE) modelling of plastic injection moulding has earned the department a CAE modelling software license donation and collaboration from Coretech Inc. He is a registered graduate mechanical engineer with the Engineers Board of Kenya.

Fredrick Madaraka Mwema

Dr Mwema is currently a Researcher (since March 2023) at Northumbria University, UK where he is working on high entropy thin film materials for extreme condition applications. He is also a Senior Lecturer, at Dedan Kimathi University of Technology (DeKUT), Kenya (currently on leave). He also served as Chair of the Department-Mechanical Engineering and Director of the Centre for Nano Materials and Nanoscience Research Centre at DeKUT (for three years since 2020). He was a lecturer (2019-2021), assistant lecturer

(2015-2019), and teaching assistant (2011-2015) of Mechanical Engineering at DeKUT. He has been a visiting student/researcher at the University of Southampton (2013), BIUST-Botswana (2018-2022), and IIT-Kharagpur, India (2018). He serves as an external examiner (since 2022) and has assessed several postgraduate theses and dissertations.

In addition, he has supervised over 100 undergraduate students, 5 master's degree and 1 doctorate students (engineering). He is currently supervising 4 PhD students and 5 master's degree students. He has published over 100 articles in peer-reviewed journals, conferences, and books (according to his Scopus profile) in his fields of interest. He has almost 1000 citations according to his Google Scholar profile. He has written four books in the areas of Thin Films, 3D printing, and Material characterization. He has successfully filed two intellectual properties (IPs) with the Kenya Industrial Property Institute (KIPI). He is a registered member of ASME, IAENG, and a graduate engineer, of the Engineers Board of Kenya (EBK).

Job Maveke Wambua
Mr. Wambua is currently a PhD student at Northumbria University, United Kingdom. He is currently working on thin film deposition of nanocomposites. He is also a tutorial fellow at the Dedan Kimathi University of Technology (DeKUT), Kenya (on study leave). He has previously worked on the processing and machining of polymeric and composite materials and has published eight articles in reputable journals and three book chapters. He is currently a graduate member of the Engineers Board of Kenya, the Institution of Engineers of Kenya, and the American Society of Mechanical Engineers.

Foreword

It is my pleasure to introduce you to the remarkable work titled "Injection Molding Process Modelling: Statistics, CAE, and AI Applications." Authored by experienced researchers and academics in materials, manufacturing, and AI, this book represents a significant contribution to the field of manufacturing and engineering.

Throughout my years in the manufacturing industry, I have witnessed firsthand the challenges and opportunities inherent in injection moulding processes. It is therefore with immense appreciation that I commend the authors for their dedication, expertise, and commitment to advancing the state-of-the-art in this field.

"Injection Molding Process Modelling: Statistics, CAE, and AI Applications" promises to be a valuable resource for professionals and practitioners alike. With its comprehensive coverage of computer-aided engineering, statistical analysis, and artificial intelligence applications, this book offers practical insights and solutions for optimizing production processes, reducing defects, and enhancing overall efficiency.

In an increasingly competitive and dynamic marketplace, the ability to design and optimize production processes is paramount. This book fills a crucial gap by providing readers with the tools, techniques, and knowledge needed to achieve excellence in injection moulding.

By leveraging the latest advancements in technology and engineering, "Injection Molding Process Modelling: Statistics, CAE, and AI Applications" has the potential to revolutionize the way industries approach production optimization. Its insights and methodologies will undoubtedly go a long way in supporting industries in their quest for cost savings, quality improvement, and operational excellence.

I extend my heartfelt congratulations to the authors for their outstanding achievements and commend them for their invaluable contributions to the field of injection moulding. It is my sincere hope that this book will inspire and empower readers to embrace innovation, drive efficiency, and achieve remarkable success in their endeavours.

With deepest admiration for the authors' dedication and expertise, I commend this book to all who seek to explore and embrace the boundless potential of the injection moulding process.

John Kimani Njagi
Plant Manager
Torrent Closures – Torrent East Africa Ltd, Nairobi
7th March 2024

Preface

The book is about computer-aided engineering (CAE), statistical techniques, and artificial intelligence (AI) and their applications in the modelling of plastic injection molding process. CAE is integral to the injection molding process, contributing to the entire product development cycle from initial design to the production of molds and final parts. The use of CAE enhances efficiency, accuracy, and collaboration, ultimately leading to better-quality products and reduced time-to-market. Specifically, CAE contributes to the injection molding in the following ways:

(a) **Product Design and Visualization**: CAE enables designers to create detailed 3D models of the product to be manufactured using injection molding. Visualization tools help in assessing the design from various angles, identifying potential issues, and making necessary modifications before moving to the manufacturing phase.

(b) **Prototyping and Testing**: CAE allows for the rapid creation of prototypes, which can be 3D printed or otherwise produced for testing purposes. Prototyping helps in evaluating the functionality, form, and fit of the product, enabling designers to adjust as needed before investing in expensive molds.

(c) **Mold Design**: CAE is essential in the design of injection molds. Mold designers use CAD software to create detailed and precise 3D models of the molds. The software helps in determining the optimal placement of injection points, designing cooling channels, and ensuring that the mold will produce parts with the desired specifications. CAD software assists in generating toolpaths for CNC machines that will be used to produce the injection molds. Toolpaths guide the cutting tools to precisely shape the mold components, ensuring accuracy and repeatability in the manufacturing process.

(d) **Analysis and Simulation**: CAE tools often include simulation capabilities that help in predicting how a design will behave during the injection molding process. Analysis features can identify potential issues such as warping, sink marks, and material flow problems, allowing designers to make informed decisions to optimize the design.

Statistical tools play a crucial role in evaluating, optimizing and controlling the injection molding process. These tools help manufacturers monitor and improve various aspects of the process, ensuring consistent product quality. Some of these tools include statistical process control (SPC), Design of Experiments (DOE), Regression Analysis, Capability Indices, Histograms and Pareto Analysis, Root Cause Analysis (RCA), and Process Failure Mode and Effects Analysis (PFMEA). By integrating these statistical tools into the injection molding process, manufacturers can enhance product quality, reduce waste, and optimize efficiency. Continuous monitoring and improvement based on statistical analysis contribute to a more robust and reliable manufacturing process.

Artificial Intelligence (AI) is increasingly being applied to various manufacturing processes, including injection molding. The integration of AI in injection molding brings the benefits of increased efficiency, improved product quality, reduced waste, and enhanced overall process control. As technology continues to advance, AI is likely to play an even more significant role in optimizing injection molding operations. Artificial intelligence finds application in injection molding in the following ways predictive maintenance, process optimization, quality control, automated decision-making, fault detection and correction, energy efficiency, material selection and optimization, supply chain optimization, and collaborative robots (Cobots) among other applications.

The landscape of manufacturing is undergoing a profound transformation, and at the heart of this revolution lies the synergy between Artificial Intelligence (AI), Computer-Aided Design (CAD), and advanced Statistical methods. In the intricate world of injection molding, where precision meets production, the interconnection of these technologies is reshaping how we conceive, design, and optimize the manufacturing process.

This book is an exploration into the dynamic realm where AI algorithms, CAE modelling, and statistical tools converge to redefine the possibilities and efficiencies in injection molding. Through this book, we journey with you through the intricacies of product design, mold creation, and the entire manufacturing workflow, uncovering how these technologies work in harmony to enhance quality control and defect reduction during the injection molding process.

The book explores the following key themes:

- Basics of injection molding as a plastic processing method and strategies to major product defect control.
- Computer-aided modelling of injection molding process with demonstrations on product design, creation, and simulation.
- Statistical tools and the application of design of experiments, parameter screening, interaction analysis, and process optimization in defect/quality control during injection molding.
- AI-based predictive modelling of injection molding for quality and defect control.

As we navigate the intricacies of these technologies, we aim to provide insights, practical applications, and a vision for the future where CAE, statistics, and AI harmonize to redefine what's possible in injection molding. Whether you are a seasoned professional in the field or a curious newcomer, this book invites you to explore cutting-edge manufacturing technology and envision the possibilities that lie at the intersection of CAE, statistics, and AI in injection molding.

Let the journey begin.

Tien-Chien
University of Johannesburg, South Africa

Edwell Tafara
Dedan Kimathi University, Kenya

Steven Otieno
Dedan Kimathi University, Kenya

Fredrick Mwema
Northumbria University, UK & University of Johannesburg, South Africa

Job Wambua
Northumbria University, UK

1

Injection Molding for Plastic Processing

1.1 Plastic Processing Methods

Plastic materials generally form the biggest percentage of domestic and industrial products due to various factors such as ease of production as a result of lower melting temperatures, lower weight, lower costs and resistance to corrosion and other forms of chemical action [1]. Plastics engineering is thus an important driver of global manufacturing scale. By providing the various means and methods to shape and form plastic materials into a variety of products, plastic processing plays a key role in material manufacturing. Figure 1.1 illustrates some of the main methods of plastic products processing categorized into extrusion, molding, additive manufacturing, and machining.

Injection molding entails the delivery of molten plastic into a mold cavity whose shape is similar to the desired product shape. Other commonly used molding methods include blow molding and thermoforming. Blow molding is geared towards the production of hollow objects and involves the use of compressed air to inflate molten plastic within a mold. The process is very popular in the packaging industry due to its suitability for high-volume production. In thermoforming, a plastic sheet is heated until it becomes pliable and then formed over a mold. This is achieved either through vacuum forming or pressure forming.

In extrusion, plastic enriched with additives is melted and forced through a die thereby creating products of continuous profiles. It is highly regarded due to its ability to produce lengthy and uniform shaped products suitable mainly for construction [2]. Additive manufacturing through fused deposition modeling entails layer by layer addition of material to create a three-dimensional product. This process has gained significant attention due to its ability to form geometrically complex products.

Of the plastic processing methods, injection molding is the most widely used technique due to various benefits associated with it such as cost effectiveness as a result of high production rate and its ability to form complex, intricate and geometrically accurate parts in large scale among others.

DOI: 10.1201/9781003492498-1

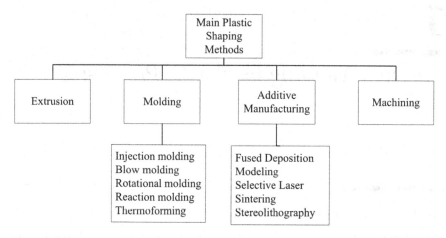

FIGURE 1.1
Main methods of plastic processing.

Injection molding has since been of great interest globally with several researches carried out in attempts to optimize and improve the performance of the production process alongside modification and improvement of the technology and equipment involved [3].

1.2 Injection Molding

Plastic Injection Molding (PIM) is a net plastic manufacturing process which involves formation of plastic components through heating a raw material to melt and delivering the molten material through pressure into a cavity with a shape similar to that of the desired product. The molten material is then held inside the cavity to cool and solidify. An injection molding machine has three main units that is the injection unit, injection mold unit and clamping unit. Figure 1.2 shows the main parts of an injection molding machine [4].

A typical plastic injection molding process comprises of four main stages which include filling, packing, cooling and ejection. During these stages, the material experiences complex thermomechanical changes which result into changes in various material properties. Inside the mold, the part is constrained within the mold surfaces and therefore residual stresses develop during solidification. These stresses are attributed to the viscoelastic flow of the polymer melt during the filling and packing stage and the solidification of the polymer melt under pressure during the cooling stage [5]. When the part is ejected, there will be stress relaxation which causes defects such as shrinkage of the part. When the different regions of the part shrink non-uniformly,

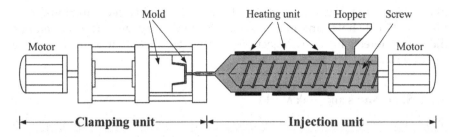

FIGURE 1.2
Main parts of an injection molding machine. (Obtained from [4] with permission from Springer Nature Ltd.)

other defects such as part warpage occurs. Shrinkage is a time dependent function and the degree depends on the molding conditions such as the temperature and pressure history of the material as it cools.

An injection mold is thus the most important equipment in the injection molding process as it forms and determines the qualities of the molded product. The mold itself is a very complex system comprising of various components subjected to cycles of stress and temperatures. Major critical features of injection mold include feed system, cooling system, ejection system and cavity layout. The main objective of feed system include conveyance of the polymer melt from the machine nozzle to the cavities while that of cooling system include enhancing polymer melt cooling and solidification. In the design and optimization of the performance of injection molding systems, numerous research has been carried out on the design for elimination of product defects [6].

Owing to vast areas of application for plastic injection molded products, quality control has become an essential aspect in the plastic injection molding process with several studies employing majorly three forms of control that is product control, machine control and process control [7]. The injection molding process quality has a direct impact on the molded product which is manifested in the form of part weight, surface integrity, structural integrity and dimensional stability.

1.3 Injection Molded Product Defects

Despite the effectiveness of plastic injection molding process in forming intricate and precise components, the components may be subject to various defects. Design for assembly guidelines requires strict conformity of the product to the tight tolerances and dimensional requirements provided by part designers. A deviation in the part dimensions will affect functionality of the part. Short shots, sink marks, shrinkage and warpage defects are among the major defects in plastic injection molded products that result to dimensional variation.

1.3.1 Short Shots and Sink Marks

Short shot defects results to incomplete cavity filling thereby forming a product that does not conform to the specified dimensional requirements as illustrated in Figure 1.3. It is a major concern in plastic injection molding as it renders the molded part completely unusable. This defect can be visually detected and signs often include incomplete mold cavity filling and variations in dimensions of the part sections.

These defects are largely caused by polymer material properties such as viscosity which affects the polymer flow. High viscosity induces flow resistance thereby requiring higher melting temperature and injection pressure to fill the cavity.

Sink marks on the other hand manifests as depressions on the surface of the molded part thereby resulting to a visually unappealing product. This majorly occurs during the cooling phase of the molding process and results from uneven shrinkage as the polymer melt solidifies. The polymer melt solidifies and undergoes volumetric contraction during the cooling phase. Rapid cooling rate leads to uneven solidification thereby causing shrinkage variations which results to sink mark variation.

1.3.2 Shrinkage and Warpage

Shrinkage refers to a geometric reduction in the molded part size and may be uniform or non-uniform. Uniform shrinkage results to a decrease in part

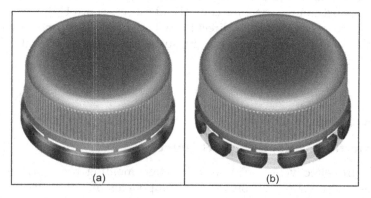

FIGURE 1.3
Illustration of a completely filled part (a) and incompletely filled part (b).

size while non-uniform shrinkage leads to warpage which is deformation or change in shape of the molded plastic part. These defects largely affects the performance, aesthetic appearance and assembly of the plastic part. Shrinkage of plastics is inevitable to a lesser extent as a result of the physics of plastic thermal contraction, plastic compressibility and thermal expansion of the mold metal [8].

During plastic injection molding process, the polymer is subjected to a series of pressure and temperature changes which vary during the various stages of filling, packing, cooling and ejection and therefore influences the state and quality of the molded product [9]. These changes in the specific volume can be utilized directly in calculation of the product linear and volumetric shrinkages. An accurate estimation of the volumetric shrinkage can be determined from the combined gas law formula with the knowledge of the pressure and temperature at the end of packing. Figure 1.4 illustrates the changes in temperature, pressure and specific volume undergone during a typical injection molding process.

While shrinkage describes in-plane dimensional changes, warpage defines out of plane dimensional changes. Warpage is caused by differential shrinkage resulting from temperature gradient through the plastic part wall. In non-uniform cooling systems, the temperature of the part in contact with the cavity insert differs from the temperature of the part in contact with the core insert. This difference leads to a difference in cooling rates and shrinkage rates. The difference in shrinkage rates on the core surface and cavity surface results to strain on the part as it cools leading to warpage [10].

Figure 1.5 illustrates the occurrence of warpage as a result of non-uniform temperature induced by non-uniform cooling system design [10]. Unbalanced cooling may not be eliminated in totality for complex shapes where the core and cavity surfaces of the part are dissimilar or have different shapes and surface areas. Differential cooling results from the unbalanced cooling thereby inducing part warpage as illustrated on Figure 1.5 [11].

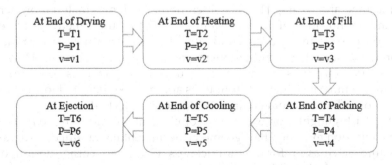

FIGURE 1.4
Temperature, specific volume and pressure changes undergone during injection molding.

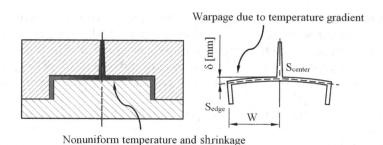

FIGURE 1.5
Warpage induced by differential cooling. (Reused from [11] under open access policy.)

Warpage and shrinkage defects are normally compensated for at the mold design stage. To compensate for volumetric shrinkage, the cavity volume is always increased by 1 - 3% whereas compensation for warpage sometimes involves reverse biasing of mold surfaces. However, larger degrees of the defects than anticipated still arises in the actual molding environment hence the need for further control. Therefore, various studies attribute the causes of shrinkage and warpage to temperature differences in the plastic part resulting from uneven contraction and uneven cooling, edge effect and molecular flow orientation [4, 12].

1.4 Strategies to Molded Product Defects Reduction

The intricate interaction between part geometry, injection mold design, polymer material properties and process parameters significantly influences the final product outcome. Major strategies to reduction of defects in plastic injection molded parts include proper design of the product through modification of the part geometry [13], design optimization of the injection molding machine and tooling [14], consideration of polymer material properties [15] and optimization of process parameters [16]. Of these methods, material selection, part geometry design and mold design are normally determined at the initial design stages and are therefore not subject to changes. However, process parameters such as temperatures and pressures are normally dynamic in most molding environments and may be subjected to changes as a result of influences from environmental conditions [13].

Table 1.1 highlights some of the major elements contributing to defects in plastic injection molding. Part geometry design entails careful consideration of features such as uniform wall thickness, rib design, corner design, draft and undercuts design. Whereas material, part geometry and mold design factors can be successfully mitigated and held constant, optimized processing

TABLE 1.1

Some of the Major Contributing Factors to Warpage and Shrinkage Defects

Mold Design	Processing Conditions	Environmental Conditions
• Feed system design	• Mold, Melt Temperatures	• Ambient Temperature
• Gate location, dimensions, number	• Injection Rate	• Humidity
	• Injection, Packing Pressures	• Air movement
• Ejection system design		
• Cooling system configuration	• Injection, Packing, Cooling, Mold Open Times	
	• Clamping Pressure	

Part Geometry	Material
• Wall thickness	• Physical Properties
• Wall thickness variations	• Chemical Properties
• Shape	• Molecular Structure
• Stiffness	• Composition

conditions may not be held constant as a result of interference from environmental conditions. Despite greater efforts to mitigate various plastic injection molded product defects through optimization of process parameters, the defects still arises as a result of interference of the molding process with several related and independent process parameters [4].

Owing to the complexity and high non-linearity between the process variables and quality parameters, advanced data driven optimization models are employed in the plastic injection molding process [17]. Application of Artificial Intelligence (AI) and Machine Learning (ML) to manufacturing processes such as injection molding is therefore a theme of current studies and provides a yardstick to quality control in injection molded products. Commonly used intelligent algorithms for defect prediction and control include artificial neural networks (ANN), genetic algorithm (GA), response surface methodology (RSM), Support Vector Machines and Kriging model [4]. Through integration of design of experiments (DOE), Computer Aided Engineering (CAE) numerical simulations and intelligent algorithms, various defects in injection molded parts have been studied, predicted and controlled resulting to efficient production and reduced scrap rates.

1.5 Conclusion

Plastic injection molding stands out as the most widely used and adaptable technique for plastic processing and its ability to develop intricate and precise parts has revolutionized the manufacturing environment. However, the inevitable occurrence of plastic injection molded product defects highlights the complex difficulties that manufacturers confront in their pursuit for quality products. This has been explored as a complex relationship between product defects and other features such as product geometry, injection mold design, polymer material characteristics and process variables.

Studies on effects of part geometry to product defects have shed light on the importance of uniform wall thickness, consideration of corners and edges and potential implications that arises from undercuts and other part features. Understanding these contributions is important to achieving a balance between the required design intricacy specified by the customer or required for product functionality and manufacturability of the part. On the effects of mold design to the product defects, studies have outlined crucial interactions among cavity layout and configuration, feed system design, cooling system design and ejection system design on product quality.

Polymer material properties and process parameters adds other layers of complexity to the quest for defect-free injection molded products. It is crucial that manufacturers strike a balance between material properties and processing conditions to prevent defects associated with incomplete filling, dimensional inaccuracies and surface imperfections. Therefore, predicting and reducing defects throughout the molding process requires an understanding of the selected polymer material behavior at given processing conditions.

Positive developments to the solution of molded product defects have entailed the modelling of the injection molding process through Computer Aided Engineering (CAE), statistical methods and intelligent data-driven models. Prior to the actual molding process, manufacturers could identify and address such problems virtually thanks to advanced mold simulation technologies through CAE. This strategy has helped to improve the overall efficiency of the production process while also saving time and resources. Furthermore, through statistical and intelligent modelling, real-time insights into the critical variables have been made possible by the development of complex process monitoring and control systems which have allowed for quick adjustments and reliable product quality. With the use of these technologies, manufacturers can keep a close watch on the injection molding procedure while swiftly resolving deviations and controlling product defects.

Therefore, while injection molding process dominates the plastic processing sector, the journey towards defect-free products remains a continuous evolution. The complex relationship between part geometry, mold design, polymer material properties and processing conditions requires constant attention through adoption of innovative technologies such as CAE modelling, statistical modelling and intelligent modelling. Through a holistic understanding of the process factors and implementation of the advanced technologies, manufacturers could solve the challenges faced in plastic injection molding and realize the full process potential for quality products.

References

[1] A. Shrivastava, "Plastic properties and testing," in *Introduction to plastics engineering*, Cambridge: William Andrew Publishing, 2018, pp. 49–110. doi: 10.1016/b978-0-323-39500-7.00003-4

[2] C. Rauwendaal, *Polymer extrusion*, 5th ed. Munich: Hanser, 2014. doi: 10.3139/9781569905395

[3] M. Czepiel, M. Bańkosz, and A. Sobczak-Kupiec, "Advanced injection molding methods: Review," *Materials*, vol. 16, no. 17, p. 5802, Aug. 24, 2023. doi: 10.3390/ma16175802

[4] N. Y. Zhao, J. Y. Lian, P. F. Wang, and Z. Bin Xu, "Recent progress in minimizing the warpage and shrinkage deformations by the optimization of process parameters in plastic injection molding: A review," *Int. J. Adv. Manuf. Technol.*, vol. 120, no. 1–2, pp. 85–101, Feb. 10, 2022. doi: 10.1007/s00170-022-08859-0

[5] M. R. Kamal, R. A. Lai-Fook, and J. R. Hernandez-Aguilar, "Residual thermal stresses in injection moldings of thermoplastics: A theoretical and experimental study," *Polym. Eng. Sci.*, vol. 42, no. 5, pp. 1098–1114, 2002. doi: 10.1002/pen.11015

[6] J. Gim and L. S. Turng, "A review of current advancements in high surface quality injection molding: Measurement, influencing factors, prediction, and control," *Polym. Test.*, vol. 115, p. 107718, Nov. 01, 2022. doi: 10.1016/j.polymertesting.2022.107718

[7] G. Gordon, D. O. Kazmer, X. Tang, Z. Fan, and R. X. Gao, "Quality control using a multivariate injection molding sensor," *Int. J. Adv. Manuf. Technol.*, vol. 78, no. 9–12, pp. 1381–1391, Jun. 2015. doi: 10.1007/s00170-014-6706-6

[8] D. O. Kazmer, *Injection mold design engineering*, 2nd ed, Munich: Hanser Publications, 2016. doi: 10.3139/9783446434196

[9] J. Wang and Q. Mao, "A novel process control methodology based on the PVT behavior of polymer for injection molding," *Adv. Polym. Technol.*, vol. 32, no. SUPPL.1, Mar. 2013. doi: 10.1002/adv.21294

[10] D. O. Kazmer, *Injection mold design engineering*, 1st ed, Munich: Hanser Publications, 2007. doi: 10.3139/9783446434196.fm

[11] A. Torres-Alba, J. M. Mercado-Colmenero, J. de D. Caballero-Garcia, and C. Martin-Doñate, "Application of new conformal cooling layouts to the green injection molding of complex slender polymeric parts with high dimensional specifications," *Polymers (Basel)*, vol. 15, no. 3, p. 558, Jan. 2023. doi: 10.3390/polym15030558

[12] J. Zhuang, D. M. Wu, H. Xu, Y. Huang, Y. Liu, and J. Y. Sun, "Edge effect in hot embossing and its influence on global pattern replication of polymer-based microneedles," *Int. Polym. Process.*, vol. 34, no. 2, pp. 231–238, Apr. 2019. doi: 10.3139/217.3726/MACHINEREADABLECITATION/RIS

[13] J. M. Fischer, "Causes of molded part variation: Processing," in *Handbook of molded part shrinkage and warpage*, 3rd ed, New York: William Andrew, 2013, pp. 81–100. doi: 10.1016/b978-1-4557-2597-7.00001-x

[14] J. Fu and Y. Ma, "Mold modification methods to fix warpage problems for plastic molding products," *Comput. Aided. Des. Appl.*, vol. 13, no. 1, pp. 138–151, Jan. 2016. doi: 10.1080/16864360.2015.1059203

[15] M. Huszar, F. Belblidia, H. M. Davies, C. Arnold, D. Bould, and J. Sienz, "Sustainable injection moulding: The impact of materials selection and gate location on part warpage and injection pressure," *Sustain. Mater. Technol.*, vol. 5, pp. 1–8, Sep. 2015. doi: 10.1016/j.susmat.2015.07.001

[16] S. Sudsawat and W. Sriseubsai, "Warpage reduction through optimized process parameters and annealed process of injection-molded plastic parts," *J. Mech. Sci. Technol.*, vol. 32, no. 10, pp. 4787–4799, Oct. 2018. doi: 10.1007/S12206-018-0926-X

[17] D. V. Rosato and M. G. Rosato, *Injection molding handbook*. Boston: Springer Science & Business Media, 2012.

2

Computer Aided Modelling for Plastic Injection Molding

2.1 Introduction

Computer Aided Engineering (CAE) modeling which incorporates Computer Aided Design (CAD) and Computer Aided Manufacturing (CAM) has revolutionized manufacturing processes as it has enabled designers and engineers to simulate, analyze, and optimize various aspects of the product manufacturing process. It has become an indispensable part of modern manufacturing and has driven advancements in various fields such as plastic processing through injection molding. Plastic injection molding process is one of the manufacturing processes that can be accurately simulated and predicted by various CAE packages.

Major CAE software packages such as Moldex3D, Autodesk Moldflow, Cadmould, Ansys and Solidworks Plastics among others have necessitated modelling and simulation of plastic injection molding process for optimization and control. These software packages have capabilities of predicting the flow patterns of the molten plastic from plastication all the way to ejection of the solidified part at the end of the molding cycle. The packages simulates the plastic flow inside the injection mold by predicting the property changes undergone by the polymer melt thereby necessitating optimum mold design and process parameter setting for quality production.

CAE modelling have been used to determine the effects of process parameters on injection molding of ultra-thin wall plastic parts [1], to determine the effects of process parameters in the molding process of a polyethylene natural fiber composite [2], in the study of warpage and shrinkage in a composite thin walled part in plastic injection molding [3, 4], for optimization of process parameters in manufacturing of top cap of water meter [5], for warpage analysis and optimization in thin walled injection molding parts [6–9], for a study on process simulation and parameter optimization of the injection molding of an automotive back door panel [10], for optimization of volumetric shrinkage and fill time for a polypropylene sample [11] and for optimization of process parameters in the control of short shot, warpage, sink marks and weldline defects [12].

Numerical simulations using CAE softwares have become popular especially for simulating the occurrence of defects such as sink marks, weldlines, short shots, shrinkage and warpage which have proven difficult to measure and quantify experimentally. The softwares are capable of quantifying the defects at each location of the part therefore giving information about the regions with the maximum values of the defects for control and optimization.

2.2 CAE Modelling of a Packaging Bottle Cap

The substantive interaction between the product design, mold design and the injection molding process calls for an iterative mold development process [13]. For this reason, product design and mold design are oftenly carried out concurrently. For this study, a two cavity three-plate plastic injection mold used for production of packaging bottle caps was developed and used as a test specimen in the CAE modelling. Moldex3D® CAE Package was utilized for defects modelling through CAE.

2.2.1 Product Design

As the study centralized on determination of the effects of processing parameters to major product defects, a packaging bottle cap was chosen as a test specimen due to its complex shape as it has various sections of thinner cross sections and literature reports that variation of process parameters has more effect to major defects for thinner sections therefore require critical care [14]. In addition, packaging caps are widely used in many applications and therefore is a justified representative of Plastic Injection Molded (PIM) products. With balanced feed systems and cooling systems, multi cavity molds of any given number of cavities will have the same responses under similar changes in process parameters. That is, a given variation in process parameters would affect the shrinkage in both a two cavity and a thirty-two cavity injection molds in the same manner. Therefore, a two-cavity mold was selected for this study for reasons such as to achieve balanced melt fill in addition to being the most commonly used mold type for low volume and complex part productions. A two cavity mold was used to generate an ANN based predictive model for warpage and shrinkage [15], for warpage and shrinkage optimization in water meter top cap manufacturing [16] and in warpage and temperature distribution optimization [9].

A CAD model of the product was developed and various aspects such as uniform wall thickness, rib design, corner design, undercuts and draft considered. Uniform wall thickness in the various regions was maintained in the bottle cap design as non-uniform wall thickness results to differential cooling which induces warpage defect. Thicker wall sections will have a longer cooling time than thinner sections therefore upon ejection, the part will exhibit higher temperatures near the thicker wall and lower temperatures near the thinner wall leading to significant part distortion [13]. Figure 2.1 shows the designed product with all the design considerations made. The largest product diameter was 30 mm, longest height 12 mm with the maximum thickness of 2.03 mm and the minimum thickness being 0.03 mm.

Internal and external ribs were designed to enhance part support and provide grip during cap opening and closing respectively. The rib design was guided by the standard requirement of a rib thickness of at least 50% of the nominal part wall thickness. Furthermore, the product design entailed avoidance of sharp corners for reasons such as, sharp corners result to regions of stress concentration which could induce warpage due to region collapse and also sharp corners may be difficult to produce in mold making. A draft angle

FIGURE 2.1
Cap CAD model.

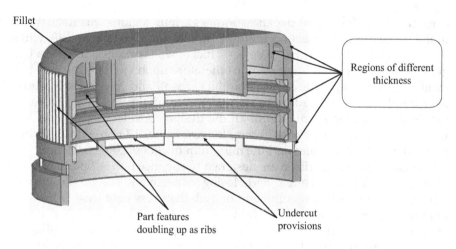

FIGURE 2.2
Product complexity.

TABLE 2.1

Polymer Material Properties

Family Name	HDPE
Grade Name	HDPE 2001 PBK44
Producer	Total Chemicals
Processing Temperatures	190–250°C
Solid Density	0.954 g/cm³
Yield Strength	24 MPa
Flexural Modulus	1.0 GPa
Poisson's Ration	0.4

of 0.5° was applied to the part to facilitate ease in part ejection. Figure 2.2 highlights the various complex features of the designed part that include regions of different wall thickness, ribs, fillets and provisions for undercuts.

The selected cap product material for this study was a commercially available grade of high density polyethylene produced by Total Chemicals under the commercial name HDPE 2001 PBK44. This material is commonly used for making plastic packaging products and some of its major properties are highlighted in Table 2.1 based on Moldex3D Material Library.

2.2.2 Mold Design

Upon product design, the design and specification of an injection mold used for manufacturing the product is made. An injection mold is the most important equipment in the injection molding process as it forms and determines the qualities of the molded product. The mold itself is a very complex system comprising of various components subjected to cycles of stress and temperatures. In the design and optimization of the performance of injection molds, lots of research has been carried out on the design of the two main injection mold systems which are the feeding system and cooling system.

For modelling of the plastic melt flow through CAE, specification of three of the most important sections of an injection mold was carried out. These sections included definition and specification of the core and cavity plates, feed system and cooling system. A two cavity three plate mold was developed for this study. The mold design process commenced with cavity definition, design and filling analysis. Cavity definition and filling analysis were carried out on Moldex3D software to determine whether the cavity would be filled in entirety during the injection molding process to eliminate possible defects which could arise due to under filling or over filling. Under filling would result to defects such as short shots while over filling would result to flash among other defects.

Feed system design entailed specification and sizing of the components of the feed system including the gates, runners and sprue to achieve the objective of conveyance of the polymer melt from the machine to the cavities. Full round cross section sprue, runners and gates were used in this study due to their excellent performance in terms of reduced pressure drop due to smoother melt flow, reduced scrap rates and increased design flexibility [13]. Their major limitation is that they have to be cut into both plates and thus require proper mold plates alignment. Other shapes such as parabolic, trapezoid and elliptical were not considered in this study for reasons such increased flow resistance, higher shear stress, uneven flow distribution and increased scrap rate [13].

A gating suitability test was carried out and yielded the best and worst regions for gate location depending on the part shape and features. This was iteratively computed using the software and based on the percentage flow length ratios. For even distribution of flow into the cavity with the lowest flow length ratio, the best recommended positions for gate location were towards the center of the top surface or bottom surface of the part as shown on Figure 2.3. The central surface region of the part had a zero flow length ratio whereas the rest of the regions had flow length ratios of between 3.4 and 50.

Minimum gate diameter was determined based on Equation (2.1) [17].

$$d = C \times N \times \sqrt[4]{A} \tag{2.1}$$

Where A is the product's cumulative surface area defined in mm². C and N are empirical constants which depend on material properties. For an average wall thickness of 2.0 mm and HDPE material, the value of C is 0.206 and N is 0.6 [17]. A total product surface area obtained from the mass properties in the

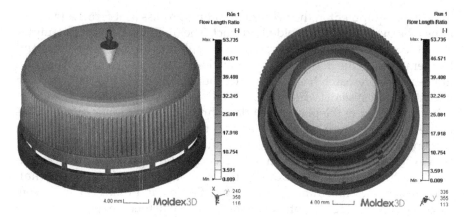

FIGURE 2.3
Gating suitability test.

3D modelling software was 4974.56 mm². A minimum possible gate diameter was found to be 1 mm. The gating configuration selected was a 3:1 which is a commonly used configuration. In this configuration, the gate diameter decreases from the runner to the cavity in the ratio 3:1. With the minimum gate diameter found to be 1 mm, the maximum gate diameter at the point of contact with the runner became 3 mm.

Figure 2.4 illustrates the gate positioning as guided by the gating suitability test.

Runner sizing was carried out based on Equation (2.2) used for the approximation of minimum runner diameters for smaller products with masses lower than 200 g and wall thickness of up to 3 mm [18].

$$D = \frac{\sqrt{w} \times \sqrt[4]{L}}{3.7} \tag{2.2}$$

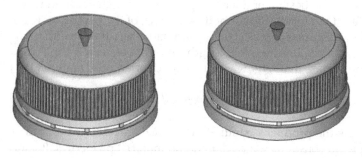

FIGURE 2.4
Gates positioning.

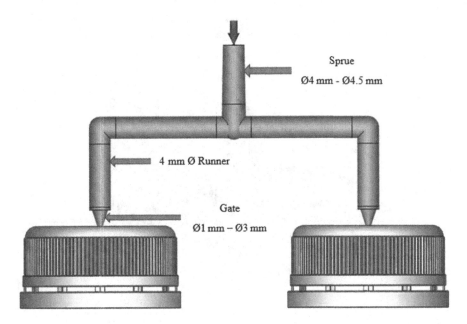

FIGURE 2.5
Mold feed system.

Where w is the total part mass in grams and L is the total runner length in mm. With a total part mass of 17 g and a total runner length of 100 mm a minimum possible runner diameter of 3.52 mm was obtained. This was within the empirical requirement that the minimum runner diameter must be 1.5 times the maximum thickness of the part [19]. Therefore, a circular runner of 4 mm diameter and sprue of 4.5 mm diameter were used for this study. Figure 2.5 shows the designed three plate feed system.

Conventional cooling system was adopted for this study. With the maximum dimensions of the cap being 30 mm diameter and 12 mm height, core and cavity plates measuring 90 mm by 90 mm by 80 mm were adopted. Eight cooling lines were adopted each with a diameter of 10 mm. Each cavity was surrounded by four cooling lines at an offset of 5 mm from the outer cavity walls to maximize heat transfer rate. Figure 2.6 illustrates the specified configuration of cooling channels used showing coolant inlet and outlet boundary conditions. Mold feed system and cooling system design validation was carried out through filling and cooling analyses to ascertain balanced fill and efficient cooling respectively.

2.2.3 Injection Molding Governing Equations

Modelling of PIM through CAE mainly involves the application of computational fluid dynamics and numerical heat transfer to develop a model that

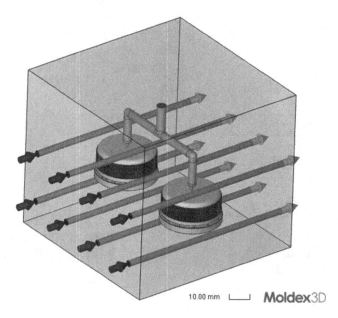

10.00 mm └──┘ **Moldex**3D

FIGURE 2.6
Cooling system.

represents the process in form of a set of complex partial differential equations and integral equations which do not have analytical solutions. Through numerical analysis, approximate and satisfactory solutions to the equations are obtained [20].

The properties of materials required for injection molding simulation depends on the product to be modelled and hence the type of simulation to be undertaken. The filling, packing and cooling analyses requires the properties of material such as viscosity, thermal sources, thermal conductivity, specific heat capacity, transition temperature and pressure-volume-temperature (PVT) data. Shrinkage and warpage analysis on the other hand requires properties such as elastic modulus, Poisson's ratios, shear modulus and coefficient of thermal expansion in addition to all the properties required in filling, packing and cooling analysis [21].

2.2.3.1 Filling Phase

The governing equations for the filling phase are typically derived from the principles of conservation of mass, momentum, and energy and are given by Equations (2.3)–(2.5) respectively as derived by Kennedy and Zheng [21] based on Hele-Shaw model simplifications.

$$\frac{D\rho}{Dt} = -\rho \nabla \cdot u \tag{2.3}$$

$$\rho \frac{\partial v}{\partial t} = \rho g - \nabla p + \nabla \cdot \eta D - \rho v \cdot \nabla v \tag{2.4}$$

$$\rho C_p \left(\frac{\partial T}{\partial t} + v \cdot \nabla T \right) = \beta T \left(\frac{\partial P}{\partial t} + v \cdot \nabla P \right) + \eta \dot{\gamma}^2 + k \nabla^2 T \tag{2.5}$$

Where ρ is the density, t is the time, u is the speed vector v is the specific volume, g is the gravitational acceleration, P is the hydrostatic pressure, Cp is the specific heat, T is the temperature, β is the heat expansion coefficient, k is the thermal conductivity and $\dot{\gamma}$ is the shear rate.

The shear rate is a function of the velocities in u, v and w directions given by Equation (2.6).

$$\dot{\gamma} = \sqrt{\left(\frac{\partial u}{\partial x} \right)^2 + \left(\frac{\partial v}{\partial y} \right)^2 + \left(\frac{\partial w}{\partial z} \right)^2} \tag{2.6}$$

The viscosity of the polymer melt was modelled as a function of the temperature, pressure and shear rate [22]. In Moldex3D software, there are a wide range of numerical viscosity models used to describe the changes in viscosity of polymers at given processing conditions. This study used the modified cross exponential viscosity model given by Equation (2.7). This model describes the dependence of shear rate across the shear thinning and upper Newtonian regions hence more suitable for thermoplastics.

$$\eta(\dot{\gamma}, T, P) = \frac{B \exp \left(\dfrac{T_b}{T} + DP \right)}{1 + \left(\dfrac{\eta_0 \dot{\gamma}}{\tau^*} \right)^{1-n}} \tag{2.7}$$

$$\eta_0(T, P) = B \exp \left(\frac{T_b}{T} + DP \right) \tag{2.8}$$

Where η_0 is the zero shear viscosity, $\dot{\gamma}$ is the effective shear rate, τ^* is the reference shear stress, n is the power law index, P is the pressure T is the temperature and the other material constants given by B, D and T_b.

The position of the polymer melt front during the molding process was determined based on a volume fraction function governed by the transport

Equation (2.9). This entailed calculation of flux at the front control volume iteratively until the cavity was filled with polymer melt.

$$\frac{\partial f}{\partial t} + \nabla \cdot (uf) = 0 \tag{2.9}$$

Where;

  $f = 0$, at the air phase
  $f = 1$, at the polymer melt phase
  $0 < f < 1$, melt front location

2.2.3.2 Packing Phase

Governing equations for the packing phase are like those of the filling phase in addition to the modified Tait Equation (2.10) used to model the material PVT behavior during the end of fill and packing phase. This model describes the behavior of specific volume at different temperatures and pressures [23].

$$V(P,T) = V(0,T)\left[1 - C \cdot \ln\left(1 + \frac{P}{B(T)}\right)\right] + V_t(P,T) \tag{2.10}$$

Where;

$$V_0(T) = \begin{cases} b_{1L} + b_{2L}\overline{T}, & T > T_t, \text{melt state} \\ b_{1s} + b_{2s}\overline{T}, & T \leq T_t, \text{solid state} \end{cases}$$

$$B(T) = \begin{cases} b_{3L}\exp\left(-b_{4L}\overline{T}\right), & T > T_t, \text{melt state} \\ b_{3s}\exp\left(-b_{4s}\overline{T}\right), & T \leq T_t, \text{solid state} \end{cases}$$

$$V_t(P,T) = \begin{cases} 0, & T > T_t, \text{melt state} \\ b_7\exp\left(b_8\overline{T} - b_9 P\right), & T \leq T_t, \text{solid state} \end{cases}$$

$\overline{T} = T - b_5$
$T_t = b_5 + b_6 P$
$b_{1L} = b_{1s}$ for amorphous polymers
$b_{1L} > b_{1s}$ for crystalline polymers
$C = 0.0894$

2.2.3.3 Cooling Phase

The cooling phase involves the consideration of a cyclic, three-dimensional and transient heat conduction problem. The three-dimensional Poisson Equation (2.11) describes the heat transfer phenomena [24].

$$\rho C_p \frac{\partial T}{\partial t} = k \left(\frac{\partial^2 T}{\partial x^2} + \frac{\partial^2 T}{\partial y^2} + \frac{\partial^2 T}{\partial z^2} \right) \tag{2.11}$$

To obtain the cycle-averaged temperature distribution of the mold base, a steady-state Laplace Equation (2.12) was solved.

$$k_m \left(\frac{\partial^2 \bar{T}}{\partial x^2} + \frac{\partial^2 \bar{T}}{\partial y^2} + \frac{\partial^2 \bar{T}}{\partial z^2} \right) = 0 \tag{2.12}$$

Where ρ is the polymer melt density, C_p is the specific heat capacity, T is the polymer melt temperature, $\bar{T}$ is the cycle-averaged mold temperature, t is the time, k is the thermal conductivity and x, y and z are Cartesian coordinates.

2.2.3.4 Warpage Phase

Standard three-dimensional residual stress models were used to model shrinkage and warpage in plastic injection molded parts. Residual stress models captures the flow-induced, pressure-induced and thermally induced stresses. The distribution of residual stresses in plastic injection molding process is determined by the varying pressure history in the packing stage and the growth of frozen layer due to cooling [21]. The models are generalizations of Hooke's law and are given by Equations (2.13) and (2.14).

$$\sigma = C \left(\varepsilon - \varepsilon^0 - \alpha \Delta T \right) \tag{2.13}$$

$$\varepsilon = \frac{1}{2} \left(\frac{\partial u_i}{\partial x_j} + \frac{\partial u_j}{\partial x_i} \right) \tag{2.14}$$

Where σ is the stress tensor, C is the stiffness tensor, ε is the strain tensor, u is the displacement tensor and α is the CLET tensor.

2.2.3.5 Solution Method

Analytical solutions to some of the shrinkage and warpage governing equations are not clearly reported in various studies. This is partly due to the complexity of the set of non-linear partial and integral governing equations and the difficulty in obtaining most of the required material data for use with the models [25]. For this reason, modelling of plastic injection molding often involves the use of numerical analysis which provides approximate but useful solutions to the modelling problems [20]. Moldex3D CAE package uses either 2.5D or 3D approximations and Finite Volume or Finite Elements for numerical solutions to the governing equations. In this study, a numerical solver based on 3D Finite Element approximation was utilized due to its suitability to solving complex geometries or irregular shapes [20], such as the one used in this study.

Continuous domains were discretized into a set of finite tetrahedral elements in three dimensional sub-domains, shape functions selected for each element and substituted into the governing equations prior to integration. The equations were multiplied by weight functions upon integration to form systems of algebraic equations in wide-area nodes. The temperature profile was determined from energy equation solution and was based on each element center.

To solve warpage phase equations, the solver considered two processes which were subjection of the part to in-mold constraints from packing to cooling phase and free deformation of the part at ejection phase where the part cooled down until thermal equilibrium with ambient air was attained. The solver used temperature profile and residual stress data as initial conditions for the solution of warpage phase equations to obtain node displacements.

2.2.3.6 Boundary Conditions

Boundary conditions were described on various surfaces such as the surface through which the melt enters the cavity, the edge of the mold, the top surface of the mold, the bottom surface of the mold, the surface defining any insert in the mold, the surface defining the melt front and the conditions describing the ejection state. These boundary conditions included [26];

The pressure gradients in the normal directions to boundaries such as mold edges and mold walls is zero meaning that the material cannot flow through or across the mold edges and walls.

$$\frac{\partial P}{\partial n} = 0 \text{ at } z = \pm h \left(\text{at the mold wall}\right) \tag{2.15}$$

$$\frac{\partial \vec{u}}{\partial z} = \frac{\partial T}{\partial z} = 0 \text{ at } z = 0 \left(\text{at the centerline}\right) \tag{2.16}$$

The melt pressure, temperature and flow rate are specified on the surface where the melt enters the cavity.

$$P_{\text{inlet}} = P\left(x, y, z, t\right);\ T_{\text{inlet}} = T_{\text{melt}} \left(\text{at the inlet}\right) \tag{2.17}$$

Taking the atmospheric pressure as the datum, the pressure at the melt front is zero as a result of mold venting.

$$P_{\text{front}} = 0 \left(\text{at the flow front}\right) \tag{2.18}$$

The temperature is uniform at the point of injection and the temperature at the mold surfaces equals the specified mold temperature.

$$\vec{u} = 0; T = T_{\text{mold}} \tag{2.19}$$

No-slip condition at the fluid-solid boundary

$$\vec{u} = 0 \left(\text{at fluid} - \text{solid boundary}\right) \tag{2.20}$$

Mass flow rate based on the total volume of part molding and the specified fill time

$$T_{\text{fill}} = 0.3\,\text{s} \tag{2.21}$$

Ejection temperature specified as recommended from the polymer material processing data sheet

$$T_{\text{eject}} = 112.35°\text{C}; T_{\text{air}} = 25°\text{C} \left(\text{at ejection}\right) \tag{2.22}$$

2.2.3.7 Model Assumptions

To simplify the computational model while maintaining its practicality, the model was based on the following assumptions [21];

I. The melt flow is considered an extended laminar flow of incompressible fluid
II. Inertial force is ignored due to high viscosity
III. Heat convection in the thickness direction is ignored
IV. Heat conduction in the direction of melt flow lower than heat convection hence ignored
V. It is possible to define physical properties such as density and viscosity at any point in the fluid and are assumed to vary smoothly over space and time.

These assumptions were made to simplify the computational model and are valid under given conditions which were the case in this study. For instance, consideration of melt flow as laminar works well with low Reynold's number flows and is the case in most injection molding processes. Turbulence in melt flow could result to variations in cooling rates which could potentially add to part warpage. However, turbulent flow patterns arises in regions of sharp melt corners and abrupt cross section area changes which were not the case in the product used for this study.

Also, ignoring inertial forces may not hold for materials of lower viscosity and higher injection speed process. However, for this study, HDPE material was used whose viscosity is very high as a result of the higher molecular weight in comparison to the other polyethylene grades. Heat transfer assumptions that were made hold for products of thinner walls which was the case in this study hence these assumptions were deemed sufficient with negligible effects on the results.

2.2.4 Finite Element Model Development

Modelling of the plastic injection molding process through CAE requires the development of the injection mold and its components as a finite element model. In this way, the complex mold structure is divided into smaller simpler elements each described by mathematical equations hence generating a mesh. Being an arrangement of elements and nodes throughout the entire system, a mesh is crucial for accurate representation and analysis.

Moldex3D uses a boundary layer mesh technology with tetrahedral mesh for injection molding simulation. Mesh geometry and refinement has significant effects on thermoplastic injection molding simulation [27]. The solver is based on an assumption of a laminar flow of a generalized Newtonian fluid. Surface mesh was generated for the product and solid mesh generated for the product, feed system, cooling system and mold base. Moldex3D allows the user to vary the mesh types for the product, feed system, cooling system and mold base. The product surface mesh transformed the product surfaces into a bunch of triangles which are used to build the geometry in virtual space during analysis.

Different styles of solid mesh were applied to the product, feed system, cooling system and mold base depending on the thickness and contribution of each section to the analysis. Inner regions of the product with the lowest wall thickness had more mesh elements specified compared to outer regions of highest wall thickness. Upon modification of node seeding and specification of the shell elements across the product surface, the boundary layer meshing technology was applied to the product solid mesh as used by [28]. Tetrahedral mesh elements with four computational nodes were used in the tetrahedral layer while prismatic elements with six computational nodes used in the boundary layers. For curve meshing, hexahedral type mesh with five inner layers and five outer layers were used for the feed system and cooling system. Tetrahedral mesh element type was used for the mold base solid mesh due to their flexibility and efficiency in capturing heat transfer which is the sole contribution of the mold base in analyses. Figure 2.7 shows the solid and surface mesh of the mold and isolated product and feed system.

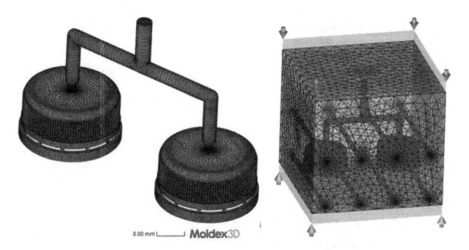

FIGURE 2.7
Finite element model of the mold.

2.2.5 Mesh Convergence Test

Prior to carrying out the numerical simulations, systematic mesh refinement was carried out on the model as illustrated on Figure 2.8. Mesh refinement was carried out to reduce the effect of mesh size and configuration on the CAE simulation results [29]. The process of mesh refinement entailed modification of the mesh parameters and carrying out numerical simulations at constant process parameters. The simulations were carried out at constant melt temperature of 200°C, mold temperature of 30°C, injection pressure of 200 MPa, packing time of 3s, packing pressure of 200 MPa and cooling lime of 8s. The initial run was carried out at default mesh properties recommended by the software. These were 3-layer BLM with a maximum surface mesh size of 2 mm and boundary layer offset ratio of 0.4. This yielded a maximum total displacement in the form of warpage of 0.20999 mm.

The mesh properties were then subsequently adjusted and the simulation repeated at the same process parameters until a convergence in the maximum total displacement achieved. The convergence was achieved beyond the mesh properties of 0.4 mm size, 5-layer BLM with offset ratio of 1. Figure 2.9 shows the results of the systematic mesh refinement process indicating the maximum total displacement obtained at the various mesh element sizes.

The difference between the results obtained at 0.8 mm, 0.4 mm and 0.2 mm was minimal. Further refinement of the mesh would just increase the computation time without significantly changing the result. Therefore, mesh properties of 5 layer BLM with maximum surface mesh size of 0.4 mm and boundary layer offset ratio of 1 were defined for the numerical simulations. 0.4 mm mesh size was used for the sections of the product with thicker walls

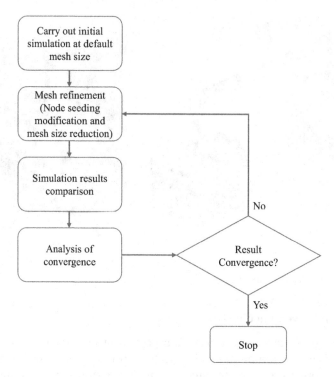

FIGURE 2.8
Mesh refinement method.

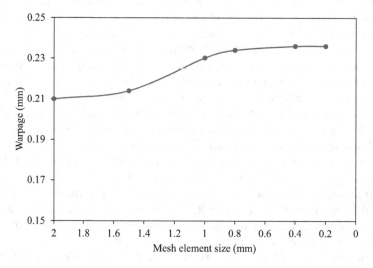

FIGURE 2.9
Mesh refinement results showing the variation of warpage with mesh size.

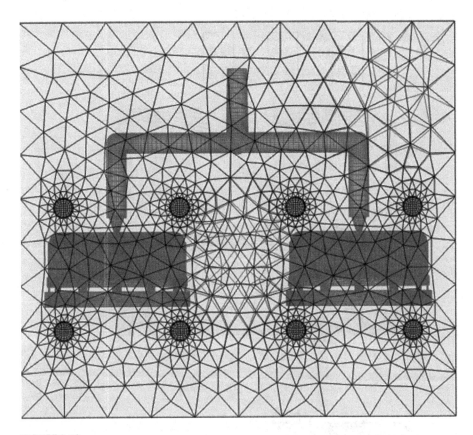

FIGURE 2.10
Finite element model of the injection mold.

and 0.2 mm size used for sections of the product with thinner walls. A five-layer BLM configuration of the part is shown in Figures 2.10 and 2.11.

2.2.6 Finite Element Model Verification

During the injection molding process, the polymer melt undergoes complex changes in specific volume and density as a result of the changes in process conditions. These changes in density affect the polymer melt flow and hence contribute to the volumetric shrinkage and warpage defects. Characterization of pressure-volume-temperature relationship is thus important in the calculation of the material compressibility during the packing phase and the final warpage and shrinkage upon ejection. Therefore, this study verified the finite element model developed through comparison of the numerical values of density obtained from the simulation against those obtained from analytical solutions at the same conditions. Analytical solutions of density were

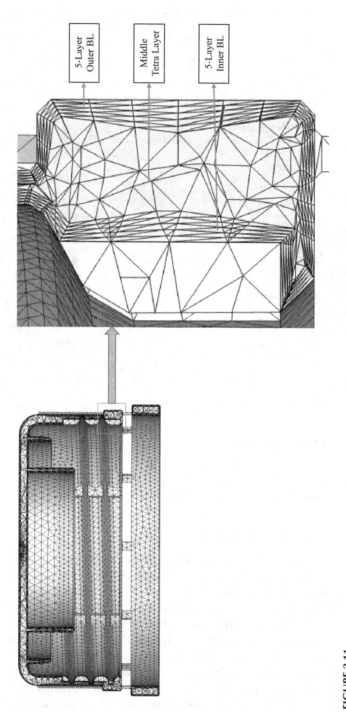

FIGURE 2.11
Part BLM configuration.

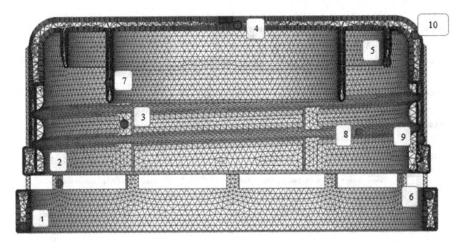

FIGURE 2.12
Sectional FE model showing the positions of the probes at various nodes.

obtained from the modified Tait Equation (2.10) which governs the change in specific volume of the polymer melt during the packing stage.

Filling and packing simulations were carried out to determine how well the solver performed calculations on the model. Ten probes were placed on various nodes of the part as shown on the Figure 2.12 where parameters such as pressure, temperature and density were obtained at specified time.

2.2.6.1 Numerical Density Results

Figure 2.13 illustrates the temperature readings taken at six of the probes at End of Fill (EOF). The temperature distributions indicated that probes 2, 3, 4 and 5 were at lower temperature at end of fill while probes 1 and 6 were at higher temperatures. This is because the polymer melt fills the part from the upper section to the lower section. Therefore, by the time the melt front reaches the node at probe 1 and 6, the other nodes would have cooled down.

Figure 2.14 illustrates the pressure distributions obtained for six of the probes. Pressure is lower on the lower region of the part with probe 1 and higher on the upper region of the part with probe 4. This is in line with the expectation as the filling profile of the part entails the flow of the melt from the upper section to the lower section. By the time the melt front reaches the lower section of the part at end of fill, the build- up of material melt makes the pressure higher at entry point and lower at the lower regions of the cavity.

Figure 2.15 illustrates the resultant density distributions obtained for the six probes. The density is lower on the nodes at higher temperatures and higher on the nodes at lower temperatures. At higher temperatures, the polymer melt expands leading to increased chain mobility and flexibility. This

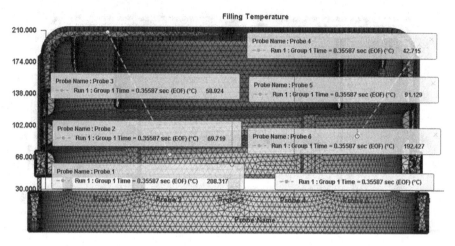

FIGURE 2.13
Temperatures at different nodes at End of Fill.

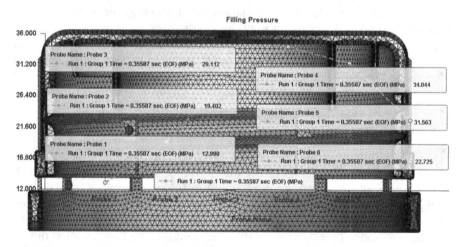

FIGURE 2.14
Pressure at different nodes at End of Fill.

makes the polymer chains easily slide past one another and occupy a larger space thereby reducing the overall density of the polymer melt.

2.2.6.2 Analytical Density Results

From the values of temperatures and the corresponding pressures at each of the probes, analytical density was computed based on the modified Tait Equation (2.10). The constants indicated in Table 2.2 are specific to Total 2001

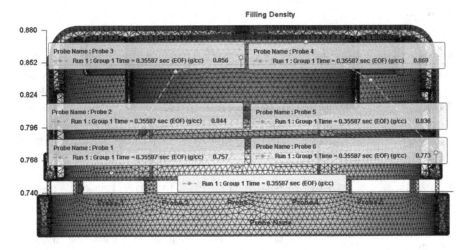

FIGURE 2.15
Densities at the various nodes at End of Fill.

TABLE 2.2

Modified Tait Equation Constants

b_{1L}	1.27E-03 m³/kg	b_{4s}	2.46E-06 1/K
b_{2L}	1.03E-06 m³/kgK	b_5	414.5 K
b_{3L}	9.26E07 Pa	b_6	2.39E-07 K/Pa
b_{4L}	4.94E-03 1/K	b_7	1.87E-04 m³/kg
b_{1s}	1.08E-03 m³/kg	b_8	5.16E-02 1/K
b_{2s}	2.08E-07 m³/kgK	b_9	1.02E-08 1/Pa
b_{3s}	3.32E08 Pa	C	0.0894

PBK 44 HDPE material and were obtained from Moldex3D result log and converted to metric units.

Parameters from the analytical equation were computed from the given values of temperature and pressure at each of the probes. Sample calculations made for six probes are illustrated on Table 2.3.

2.2.6.3 Density Comparison

Figure 2.16 shows the analytical and numerical density plots. Of the ten probes, the largest deviation between the analytical and numerical density was 0.0046 g/cm³ which was significantly lower hence satisfactory. These deviations were attributed to slight differences in underlying considerations during computations. Although analytical computations assume steady-state conditions, numerical simulation takes into account the transient behavior of

TABLE 2.3

Sample Analytical Density Computation Results

Probe	1	2	3	4	5	6
Temperature T (K)	481.47	342.87	332.07	315.87	364.28	465.58
Pressure P (Pa)	1.3E+07	1.9E+07	2.9E+07	3.4E+07	3.2E+07	2.3E+07
T_t (K)	417.60	419.14	421.46	422.64	422.04	419.93
$\bar{T}$ (K)	66.967	-71.631	-82.426	-98.635	-50.221	51.077
V_0 (m³/kg)	1.34E-03	1.20E-03	1.19E-03	1.17E-03	1.22E-03	1.32E-03
B (Pa)	6.65E+07	1.32E+08	1.39E+08	1.51E+08	1.19E+08	7.19E+07
V_t (m³/kg)	0	3.81E-06	1.98E-06	8.14E-07	1.02E-05	0
V (m³/kg)	1.32E-03	1.19E-03	1.17E-03	1.15E-03	1.20E-03	1.29E-03
Analytical Density (g/cm³)	0.7589	0.8436	0.8569	0.8711	0.8314	0.7751
Numerical Density (g/cm³)	0.757	0.844	0.856	0.869	0.836	0.773
Deviation in Density (g/cm³)	0.002	0.000	0.001	0.002	-0.005	0.002

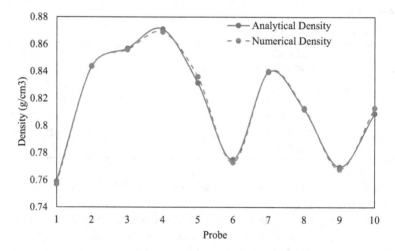

FIGURE 2.16
Density verification plots.

the polymer material properties hence the difference. The model therefore accurately predicted the changes in specific volume and hence the density of the polymer melt as a result of the changes in temperatures and pressures throughout the molding cycle. A match in the plots indicated an effective model characterization of the PVT relationship and implied model suitability to the injection molding simulations.

2.2.7 Mold Design Validation

Mold design validation entails carrying out a flow simulation analysis to determine the effect of the design to the flow characteristics of the polymer melt as it is delivered into the cavities. CAE packages discretizes the design geometry of the product into mesh of finite elements such that the temperature, pressure and flow may be calculated. The mold design validation has been used to determine the design integrity of the specified feed system and cooling system prior to further numerical simulations for defect modelling. Major areas of concern obtained and addressed through fill analysis include whether the part would fill to 100% as required, whether the flows to the cavities are balanced and whether there would be resulting packing issues.

The main objectives in cavity filling analysis include to ascertain complete filling of mold cavities, to avoid uneven filling or over-packing and to control the melt flow. Polymer flow analysis requires an understanding of the relationship between viscosity, shear rate and shear stress [30]. Shear stress measures the force per unit area exerted by the fluid as it flows, shear rate measures the rate at which the fluid layers slide over each other during flow while the viscosity measures the fluid resistance to flow and are related through the Newtonian model Equation (2.23).

$$\tau = \eta\dot{\gamma} \qquad\qquad (2.23)$$

Where τ is the shear stress, η is the viscosity and $\dot{\gamma}$ is the shear rate. Melt filling results provides an insight of whether the part gets completely filled during the filling stage of injection molding process to prevent defects such as short shot.

Preliminary mold filling analysis was carried out at an actual filling time of 0.3 seconds and a gate freeze time of 3.63 seconds. A resulting filling melt front advancement time is shown in Figure 2.17. The contour plot shows a completely colored part indicating a complete fill. The filling log was further checked to confirm a full fill percentage of 100%. From the melt front advancement, the filling pattern of the molding and the gate contributions to balanced flow was established. The uniform melt front advancement time indicated a uniform melt front velocity into the respective cavities which is recommended to minimize part non-uniformity [31].

The similarity of flow patterns between the two cavities was an indication of a balanced flow which is a major requirement in multi-cavity injections molds. This was further confirmed from the gate contribution result showing that each gate contributed exactly 50.0% to melt flow into the cavities as illustrated on Figure 2.18. Even filling allows for lower and uniform melt pressures throughout the mold cavity.

Figure 2.19 illustrates the obtained flow rate profile during the filling stage. This profile was typical of any plastic injection molding process filling

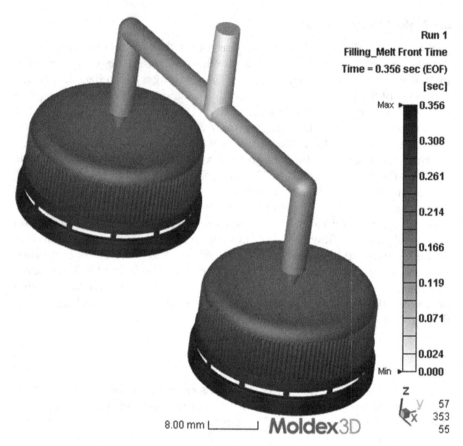

FIGURE 2.17
Melt front time.

stage where the polymer melt flow rate gradually increases during the start of fill and decreases towards the end of fill which is attributed to resistance to flow and the pressure drop across the flow front. From this profile, the first 25% of the fill time involved flow of the polymer melt along the feed system. During this period, the resistance to flow was higher and the pressure drop across the flow front lower resulting to a lower melt flow rate. A higher melt flow rate was attained when the polymer melt flows into the cavity as a result of a lower resistance to flow and higher pressure drop across the flow front. End of fill takes up 7% of the fill time where the melt flow rate drops as a result of a higher resistance to flow as the cavities are nearly or entirely filled.

Three major objectives of feed system balancing in identical part molding include to balance the fill time into the cavities, the shear rates and packing pressure within the cavities all of which were ascertained from the fill

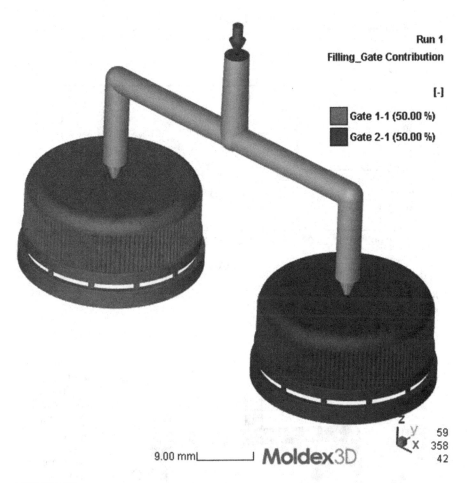

FIGURE 2.18
Gate contribution.

analysis. Figure 2.20 shows the cooling efficiencies of each of the cooling channels taken as percentages of all the cooling channels. It shows the percentage of heat withdrawn by the given cooling line with respect to the total heat and the summation of the individual contributions would equal 100%. Q_i represented the total heat that flows into the ith cooling channel surface. A positive Q_i and hence a positive cooling efficiency meant that heat was absorbed from the cooling channel surface by the cooling fluid while a negative cooling efficiency would mean that heat is released from the cooling channel surface by the cooling fluid thus indicating a heating effect instead of cooling. The filling and cooling simulations therefore confirmed the mold design integrity and suitability for use in further numerical simulations for warpage and shrinkage modelling.

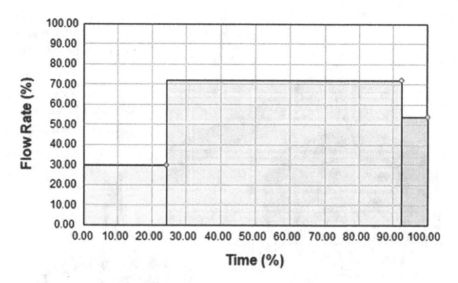

FIGURE 2.19
Obtained flow rate profile.

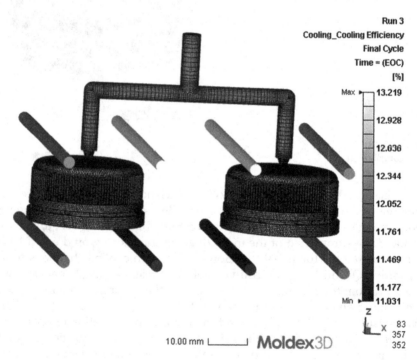

FIGURE 2.20
Channel cooling line contributions to total efficiency.

2.2.8 Process Parameter Settings

Upon carrying out literature review to determine various significant parameters influencing major defects, six process parameters including melt temperature, mold temperature, injection pressure, packing pressure, cooling time and packing time were adopted. The selection of the levels of application of each of the factors was guided by the selected material operating levels as recommended by the manufacturer.

A full analysis encompassing filling + packing + cooling + warpage was carried out. Figure 2.21 illustrates the major defects determined. Warpage was represented by total displacement obtained as the vector sum of the displacements along the x-axis, y-axis and z-axis of the molded part. The maximum total displacement and maximum shrinkage gave the regions of the part with the maximum displaced nodes and nodes of maximum size reductions respectively. Sink marks were measured in terms of displacements of the nodes from the part surface to form depressions. Short shot possibility factor was computed based on Equation (2.24) with a higher factor indicating higher chances of short shot defect [32].

$$S_f = \frac{\text{Max.Cavity Pressure}}{\text{Injection Pressure Setting}} \tag{2.24}$$

2.2.9 Numerical Data Validation

The variability of injection molding process in terms of the dependence on molding machine, injection mold, part geometry, processing conditions and environmental conditions limits quantitative comparison of defects data against those of other studies. When all the other factors are held constant, a given change in a molding process parameter affects a particular defect in the same manner irrespective of the molding environment or condition. A qualitative validation technique was used in this study and involved comparison of the main effect plots obtained from the study against those obtained experimentally and reported in literature.

Upon carrying out a full analysis which included filling, packing, cooling and warpage, the shrinkage defect mains effect plots from warpage phase were made and compared against those reported in literature. The trends in the mains effect plots were compared against those published by Mukras et al. [33], who carried out experimental studies on effects of process parameters to shrinkage and warpage defects. In their study, the authors varied seven process parameters through central composite design using HDPE polymer material grade to determine the effects of variation of process parameters to shrinkage and warpage of thin rectangular plate. The study quantified

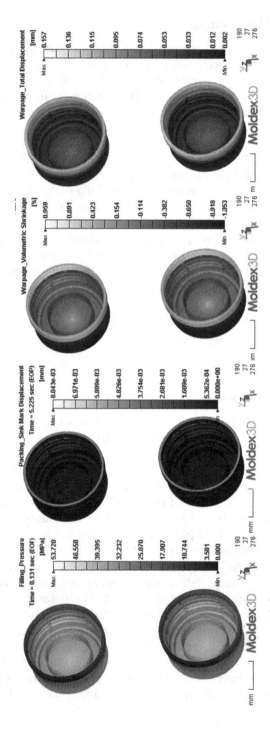

FIGURE 2.21
Contour plots showing cavity pressure, sink mark depth, volumetric shrinkage and warpage defects.

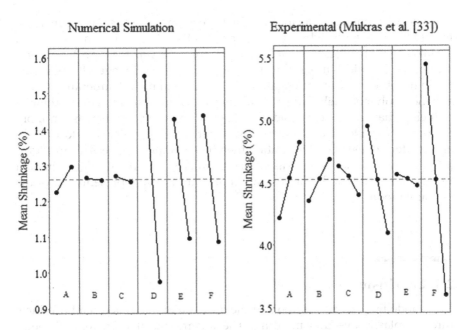

FIGURE 2.22
Comparative mains effect plots of numerical simulation and experiment data trend for *A*-melt temperature, *B*-mold temperature, *C*-injection pressure, *D*-packing pressure, *E*-cooling time and *F*-packing time.

shrinkage in terms of the change in molded product mass while warpage was quantified in terms of the sum of out of plane displacements from the product edges.

Similarity was obtained in the data trends as illustrated on Figure 2.22. A general increase in melt temperature increased shrinkage whereas an increase in injection pressure, packing pressure, packing time and cooling time decreased shrinkage. Similar trends were reported by Abasalizadeh et al. [34] who also carried out the study experimentally.

The simulation study indicates an increase in shrinkage with an increase in melt temperature which is the case as the experimental studies. This is as a result of the increase in molecular disintegration with an increase in melt temperature thereby increasing the rate of shrinkage. An increase in injection pressure ensures more material delivery and higher material compression inside the mold thereby lowering chances of volumetric shrinkage. An increase in packing time as well as packing pressure enhances an increase in material compensation within the mold thereby reducing shrinkage. Increase in cooling time ensures an increase in stress relaxation time thereby reducing shrinkage [34].

However, numerical data indicated a decrease in shrinkage with an increase in mold temperature while the experimental study reports an increase in shrinkage with increasing mold temperature. As reported by Mieth and Tromm [35], mold temperature have an effect on shrinkage effects which could be either positive or negative. An increase in mold temperature could enhance polymer melt fluidity and increase flow into the cavity thereby reducing chances of post-mold shrinkage. Conversely, an increase in mold temperature could also result to slower part cooling which would encourage greater degree of crystallization thereby resulting to increased shrinkage [36]. A match in the trends was deemed sufficient and the similarity in data trends indicated that the numerical model was a close representation of the physical model.

2.3 Conclusion

CAE modelling has revolutionized the design and manufacture of products through plastic injection molding. The integration of CAE tools through injection molding process modelling and simulation has become a valuable asset to plastic processing entities. Simulation of the polymer material behavior under varying processing conditions helps in foreseeing potential processing issues and provides insights on potential areas of modification for process improvement.

The heart of a successful CAE modelling lies in product design, mold design, mold design validation, finite element model development and verification, and numerical data validation which are fundamental in achieving accurate modelling results. As CAE modelling is just a prediction of the actual physical system, each of these processes must be correctly carried out to obtain accurate data. Mold design validation through filling analysis is used to ascertain that the melt fills the cavities at achievable molding pressures. Through cavity filling analysis, the melt front temperature, shear rates, gate contributions and fill profiles are determined to ascertain uniform and balanced fill.

The development of a finite element model of the mold is a very crucial stage in CAE modelling. Accurate modeling of the part geometry, mold geometry, material properties, and boundary conditions enhances a realistic representation of the physical injection molding process and environment. Verification of this model through extensive testing against analytical models guarantees its reliability and accuracy. Furthermore, comparison of the numerical results with physical tests validates the accuracy of the model and ensures that the predictions align with real-world outcomes.

Therefore, the utilization of CAE for modelling of plastic injection molding process coupled with a comprehensive understanding of major CAE modelling stages and requirements helps to streamline the manufacturing process and also ensures the production of high-quality, defect-free components.

References

[1] M. C. Song, Z. Liu, M. J. Wang, T. M. Yu, and D. Y. Zhao, "Research on effects of injection process parameters on the molding process for ultra-thin wall plastic parts," *J. Mater. Process. Technol.*, vol. 187–188, pp. 668–671, 2007. doi: 10.1016/j.jmatprotec.2006.11.103

[2] W. A. Rahman, L. T. Sin, and A. R. Rahmat, "Injection molding simulation analysis of natural fiber composite window frame," *J. Mater. Process. Technol.*, vol. 197, no. 1–3, pp. 22–30, 2008. doi: 10.1016/j.jmatprotec.2007.06.014

[3] Chen, M. T. Chuang, Y. H. Hsiao, Y. K. Yang, and C. H. Tsai, "Simulation and experimental study in determining injection molding process parameters for thin-shell plastic parts via design of experiments analysis," *Expert Syst. Appl.*, vol. 36, no. 7, pp. 10752–10759, 2009. doi: 10.1016/j.eswa.2009.02.017

[4] M. D. Azaman, S. M. Sapuan, S. Sulaiman, E. S. Zainudin, and A. Khalina, "Shrinkages and warpage in the processability of wood-filled polypropylene composite thin-walled parts formed by injection molding," *Mater. Des.*, vol. 52, pp. 1018–1026, 2013. doi: 10.1016/j.matdes.2013.06.047

[5] N. Sateesh, S. D. Reddy, G. P. Kumar, and R. Subbiah, "Optimization of injection molding process in manufacturing the top cap of water meter," *Mater. Today Proc.*, vol. 18, pp. 4556–4565, Jan. 2019. doi: 10.1016/J.MATPR.2019.07.430

[6] S. Sudsawat and W. Sriseubsai, "Warpage reduction through optimized process parameters and annealed process of injection-molded plastic parts," *J. Mech. Sci. Technol. 2018 3210*, vol. 32, no. 10, pp. 4787–4799, Oct. 2018. doi: 10.1007/S12206-018-0926-X

[7] K. Saeedabadi, G. Tosello, and M. Calaon, "Optimization of injection molded polymer lab-on-a-chip for acoustic blood plasma separation using virtual design of experiment," *Procedia CIRP*, vol. 107, pp. 40–45, Jan. 2022. doi: 10.1016/J.PROCIR.2022.04.007

[8] J. Chen, Y. Cui, Y. Liu, and J. Cui, "Design and parametric optimization of the injection molding process using statistical analysis and numerical simulation," *Processes*, vol. 11, no. 414, pp. 1–17, 2023. doi: 10.3390/pr11020414

[9] W. T. Huang, C. L. Tsai, W. H. Ho, and J. H. Chou, "Application of intelligent modeling method to optimize the multiple quality characteristics of the injection molding process of automobile lock parts," *Polymers (Basel)*, vol. 13, no. 15, 2021. doi: 10.3390/polym13152515

[10] Wang, Y. Wang, and D. Yang, "Study on automotive back door panel injection molding process simulation and process parameter optimization," *Adv. Mater. Sci. Eng.*, vol. 2021, p. 9996423, 2021. doi: 10.1155/2021/9996423

[11] K. Jain, D. Somwanshi, and A. Jain, "Effect of process parameter on plastic parts using ANOVA with moldflow simulation," *Adv. Mater. Process. Manuf. Appl.*, pp. 421–429, 2021. doi: 10.1007/978-981-16-0909-1_43/COVER

[12] M. Moayyedian, *Intelligent optimization of mold design and process parameters in injection molding.* 2019.

[13] D. O. Kazmer, *Injection mold design engineering*, 1st ed. Munich: Hanser Publications, 2007. doi: 10.3139/9783446434196.fm

[14] H. Oktem, T. Erzurumlu, and I. Uzman, "Application of Taguchi optimization technique in determining plastic injection molding process parameters for a thin-shell part," *Mater. Des.*, vol. 28, no. 4, pp. 1271–1278, 2007. doi: 10.1016/j.matdes.2005.12.013

[15] Z. Song, S. Liu, X. Wang, and Z. Hu, "Optimization and prediction of volume shrinkage and warpage of injection-molded thin-walled parts based on neural network," *Int. J. Adv. Manuf. Tech.*, vol. 109, no. 3–4, pp. 755–769, Jul. 2020. doi: 10.1007/s00170-020-05558-6

[16] N. Sateesh, S. D. Reddy, G. P. Kumar, and R. Subbiah, "Optimization of injection molding process in manufacturing the top cap of water meter," *Mater. Today Proc.*, vol. 18, pp. 4556–4565, 2019. doi: 10.1016/j.matpr.2019.07.430

[17] P. Jones, *The mould design guide*, 2nd ed., vol. 53, no. 9. Shrewsbury: Smithers Rapra Technology Limited, 2010. [Online]. Available: http://www.rapra.net

[18] M. Moayyedian, K. Abhary, and R. Marian, "Elliptical cross sectional shape of runner system in injection mold design," *Int. J. Plast. Technol.*, 2016. doi: 10.1007/s12588-016-9153-4

[19] J. Beaumont, *Runner and gating design handbook:Tools for successful injection molding*, 3rd ed., Hanser Publications, 2019. doi: doi.org/10.1016/C2014-0-02053-9

[20] H. Zhou, *Computer modeling for injection molding: Simulation, optimization, and control*, 1st ed., New Jersey: John Wiley and Sons Inc, 2013. doi: 10.1002/9781118444887

[21] P. Kennedy and R. Zheng, *Flow analysis of injection molds*, 2nd ed., Munich: Hanser Publications, 2013. doi: 10.3139/9781569905227

[22] W. C. Lin, F. Y. Fan, C. F. Huang, Y. K. Shen, and H. Wang, "Analysis of the warpage phenomenon of micro-sized parts with precision injection molding by experiment, numerical simulation, and grey theory," *Polymers (Basel).*, vol. 14, no. 9, 2022. doi: 10.3390/polym14091845

[23] R. Chang, C. Chen, and K. Su, "Modifying the tait equation with cooling-rate effects to predict the pressure–volume–temperature behaviors of amorphous polymers: Modeling and experiments," *Polym. Eng. Sci.*, vol. 36, no. 13, pp. 1789–1795, 1996. doi: 10.1002/pen.10574

[24] M.-L. Wang, R.-Y. Chang, and C.-H. (David) Hsu, "Molding simulation: Theory and practice," in *Molding simulation: Theory and practice*, Hanser, 2018, pp. I–XVIII. doi: 10.3139/9781569906200.fm

[25] P. Kennedy and R. Zheng, "Shrinkage of injection-molded material," in *Precision injection molding: Process, materials and applications*, J. Greener and R. Wimberger-Friedl, Eds., 1st ed., Munich: Hanser Gardner, 2006, pp. 105–135.

[26] M. Huszar, F. Belblidia, H. M. Davies, C. Arnold, D. Bould, and J. Sienz, "Sustainable injection molding: The impact of materials selection and gate location on part warpage and injection pressure," *Sustain. Mater. Technol.*, vol. 5, pp. 1–8, Sep. 2015. doi: 10.1016/j.susmat.2015.07.001

[27] D. A. De Miranda, "Influence of mesh geometry and mesh refinement on mathematical models of thermoplastic injection simulation tools," *IOSR J. Mech. Civ. Eng.*, vol. 15, no. 3, pp. 38–44, 2018. doi: 10.9790/1684-1503013844

[28] H. Zhang, F. Fang, M. D. Gilchrist, and N. Zhang, "Precision replication of micro features using micro injection molding: Process simulation and validation," *Mater. Des.*, vol. 177, no. 9, p. 107829, Sep. 2019. doi: 10.1016/j.matdes.2019.107829

[29] D. Loaldi *et al.*, "Experimental validation of injection molding simulations of 3D microparts and microstructured components using virtual design of experiments and multi-scale modeling," *Micromachines*, vol. 11, no. 6, p. 614, Jun. 2020. doi: 10.3390/MI11060614

[30] J. Koszkul and J. Nabialek, "Viscosity models in simulation of the filling stage of the injection molding process," *J. Mater. Process. Technol.*, vol. 157–158, no. SPEC. ISS., pp. 183–187, Dec. 2004. doi: 10.1016/J.JMATPROTEC.2004.09.027

[31] X. Chen and F. Gao, "Profiling of injection velocity for uniform mold filling," *Adv. Polym. Technol.*, vol. 25, no. 1, pp. 13–21, 2006. doi: 10.1002/adv.20054

[32] M. Moayyedian, K. Abhary, and R. Marian, "The analysis of short shot possibility in injection molding process," *Int. J. Adv. Manuf. Technol.*, vol. 91, no. 9–12, pp. 3977–3989, 2017. doi: 10.1007/s00170-017-0055-1

[33] S. M. S. Mukras, H. M. Omar, and F. A. Al-Mufadi, "Experimental-based multi-objective optimization of injection molding process parameters," *Arab. J. Sci. Eng.*, vol. 44, no. 9, pp. 7653–7665, 2019. doi: 10.1007/s13369-019-03855-1

[34] M. Abasalizadeh, R. Hasanzadeh, Z. Mohamadian, T. Azdast, and M. Rostami, "Experimental study to optimize shrinkage behavior of semi-crystalline and amorphous thermoplastics," *Iran. J. Mater. Sci. Eng.*, vol. 15, no. 4, pp. 41–51, 2018. doi: 10.22068/ijmse.15.4.41

[35] F. Mieth and M. Tromm, "Multicomponent technologies," in *Specialized injection molding techniques*, William Andrew Publishing, 2016, pp. 1–51. doi: 10.1016/B978-0-323-34100-4.00001-8

[36] J. M. Fischer, "Causes of molded part variation: Processing," in *Handbook of molded part shrinkage and warpage*, 3rd ed., William Andrew, 2013, pp. 81–100. doi: 10.1016/b978-1-4557-2597-7.00001-x

3

Statistical Modelling of Plastic Injection Molding Defects

3.1 Introduction

The application of statistical tools and techniques has revolutionized plastic injection molding process by enabling process analysis, prediction and optimization. Statistical models helps in understanding the relationship between various process parameters and the occurrence of defects. Through mains effects, interaction effects, and ANOVA, one can identify the optimal conditions that minimize defects and enhance overall product quality. Traditionally, a trial and error method was used to determine relationship between various injection molding parameters and relied highly on experience and expertise of operators to set the right process parameters [1].

Presently, through design of experiments, one can systematically and structurally explore and understand the complex relationships among process parameters thereby making informed decisions on parameter selection and setting. Design of experiments is the basis of prediction algorithms in cases where historical data is not available and a prediction of the defects need to be made based on a set of data. In design of experiments, various process parameters are varied in different combinations and quality indices obtained. The result of the quality indices are therefore used to optimize the process parameters depending on the given objectives. In plastic injection molding process, different design of experiment methods have been applied.

3.2 Response Surface Designs

Response Surface Methodology (RSM) applies statistical techniques to model and analyze the relationship between multiple independent variables and the output responses. In plastic injection molding, it is useful in optimization of processes by identifying the settings for input variables that yield

DOI: 10.1201/9781003492498-3

the optimal output. RSM adopts quadratic terms hence useful in curvature modelling of various responses. Central Composite Designs (CCD) and Box-Behnken Designs (BBD) are some of the main types of response surface designs applicable in plastic injection molding defects modelling.

Various studies have utilized CCD in defects modelling and study. These include optimization of process parameters for warpage and volumetric shrinkage defects control [2], warpage optimization on a plastic part [3], modelling and prediction of sink mark depth [4], analysis of warpage phenomenon of micro-sized parts [5], optimization of process parameters for flash and part weight variation control [6], optimization of process parameters for dark spots and short shot defects control [7], and optimization of process parameters for sink mark minimization [8].

Compared to CCD, BBD have fewer experimental runs but are limited to a maximum of three levels for each input parameter. Box-Behnken designs have been adopted in various studies such as experimental study of warpage and shrinkage defects [9], modelling of weld-line width and sink mark depth [10], process parameter optimization for warpage minimization [11] and optimization of process parameters for quality index improvement [12]. However, RSM designs entails many experimental runs in cases where many factors are involved and may not consider effects of interactions among variables.

3.3 Taguchi Array Designs

Taguchi design of experiment has been used by a wide range of studies owing to its capability of providing maximum amount of information from a minimum number of experimental trials [13]. Analysis of variance and signal to noise ratio are various approaches used to assess the quality of results obtained by this method of experiment designs. As a result of the higher costs of conducting experiments in plastic injection molding, a wide range of studies resorted to the use of Taguchi experiment design as a result of fewer number of experiment runs yielded [14].

Taguchi method has been used by various studies for purposes such as analyzing warpage and shrinkage properties in micro gears polymer composites [15], optimization of process parameters with an aim of minimizing warpage and shrinkage defects [16–19], optimization of process parameters for reduction of sink mark defects [20–23], process optimization to minimize weldlines [24] and in the analysis and optimization of process parameters for short shot defect control [25, 26]. Despite the many strengths of Taguchi designs with respect to determination of maximum information from minimum number of experimental runs, Taguchi designs do not give information about the effects of interactions among variables onto the response.

3.4 Factorial Designs

The commonly used experimental designs such as RSM and Taguchi designs does not give a detailed information about the effects of interaction of process parameters and how these interactions affects quality in plastic injection molding [27]. Injection molding process is a very intricate process and involves complex interactions of process parameters in affecting the various performance indices. Therefore, a full factorial design of experiment is capable of giving full detailed information about the interaction of process parameters and their effects. However, a disadvantage of full factorial design is that when the number of factors to be tested increases, the number of experiments to be carried out increases exponentially.

If higher order interactions are negligible, as is always the case in many processes, a fractional factorial design can be used. A fractional factorial design of a higher resolution is capable of providing some interaction effects in a less number of experimental runs. Generally, some three and higher order interactions are always rare and thus can be ignored. Fractional factorial design is always used in design of experiment screening for experiments involving many factors [28].

A fractional experiment design is generally given by the Equation (3.1);

$$N_{\text{fractional}} = m(n)^{k-q} \tag{3.1}$$

Where N is the number of experimental runs to be carried out, m is the number of replicates, n is the number of levels of application of each factor, k is the number of factors to be investigated and q is the degree of fractionating. Limited studies have utilized fractional factorial designs in plastic injection molding defects control. Fractional factorial designs were used to evaluate the effects of process parameters in warpage and shrinkage defects control [29, 30].

3.5 Statistical Modelling of a Packaging Bottle Cap

Alongside mains effects analysis which determines the effects of the individual process parameters to the defects, an interaction effect analysis should be carried out. Interaction effect analysis determines the effects of the interaction between process parameters to the defects. Most previous studies typically concentrated on investigating the mains effects between process parameters and quality indices and therefore overlooking interactions and their effects. Because interaction effects can either minimize or maximize individual mains effects, their determination and quantification is important. Studies report that ignoring interaction effects for some complex systems may result to chances of wrong statistical inference, missing out on important parameters and making biased predictive estimates [31].

3.5.1 Design of Experiment

Therefore, the case study determined the interactions among major process parameters and how those interactions affect short shot, sink marks, warpage and shrinkage defects in a plastic injection molded part. To establish mains effects as well as effects of process parameter interactions to major defects, a two level half-fractional factorial design of resolution VI was used for the case study. Six factors were used each at two levels as illustrated on Table 3.1. The levels were specific to Total 2001 PBK 44 HDPE material obtained from Moldex3D® material library.

The half fractional experiment design was carried out and the specified features of the design included zero center points per block, one replicate for corner points, one block and interactions up through order six yielding 32 experimental runs. CAE simulations were run at each process parameter combination to determine warpage, shrinkage, sink mark and short shot defect responses.

Mains effect plots and Pareto charts were constructed to visually represent the extent of the process parameter effects to the defects. For six process parameters, up to six-way interactions can be obtained. However, four and higher order interactions are very rare and are deemed insignificant. This study therefore considers all two-way and a few three-way interaction effects. Interaction analysis involved the determination of the significance of the interactions as well as the sizes of the interaction terms.

Mains effect and interaction effect sizes were computed based on a difference of means method [32]. By substituting the lower parameter levels with -1 and higher levels with $+1$, the mains effect and interaction effect sizes

TABLE 3.1

Selected Factorial DOE Process Parameters and Levels

Process Parameters	Levels	
	Min	Max
Melt Temperature (°C) (A)	200	235
Mold Temperature (°C) (B)	30	50
Maximum Injection Pressure (MPa) (C)	200	300
Maximum Packing Pressure (MPa) (D)	200	300
Cooling Time (s) (E)	8	15
Packing Time (s) (F)	3	5

were computed as the difference between average shrinkage and warpage responses at high (+) levels and low (−) levels of the effects as highlighted by Equation (3.2).

$$\text{Effect} = \frac{1\left(\text{Sum response at all} + \text{levels}\right)}{16} - \frac{1\left(\text{sum response at all} - \text{levels}\right)}{16} \quad (3.2)$$

3.5.2 CAE Results

Table 3.2 illustrates the resulting defects obtained at various process parameter combinations. The numerical data was validated qualitatively as described in Section 2.2.9 of Chapter 2.

3.5.3 Mains Effects

Mains effect plots determines the individual effects of the process parameters to the defects when all the other parameters are kept constant and the effects of the individual factors deemed independent.

3.5.3.1 Warpage

Figure 3.1 shows the mains effect plot for warpage. The trends indicate packing pressure as the process parameter with the highest individual effect followed by melt temperature while mold temperature, injection pressure and packing time does not largely affect warpage.

An increase in packing pressure substantially reduces warpage as a result of enhanced molecular alignment. At high packing pressure, polymer chains are oriented in an organized manner and this orientation further reduces the internal stresses within the material and helps it maintain its shape and dimension. This leads to a significant reduction in warpage defect.

An increase in melt temperature decreases warpage. Increasing the melt temperature ensures the material remains molten for a longer time during the cooling phase and thus undergoes a uniform solidification which minimizes the stresses that contribute to warpage. Similarly, a study by Chen and Zhu [33], obtained a significant decrease in warpage at an increasing packing pressure and melt temperature. Moreover, an experimental study by Singh et al. [34] obtained similar trends in terms of a decrease in warpage with increasing melt temperature, packing pressure and packing time.

3.5.3.2 Shrinkage

Main effect plot for shrinkage is shown in the Figure 3.2. An increase in cooling time resulted to a decrease in shrinkage rate. Increased cooling time

TABLE 3.2

Numerical Modelling Results

Run	A	B	C	D	E	F	Warpage (mm)	Shrinkage (%)	Short shot ratio	Sink mark (mm)
1	−1	−1	+1	−1	−1	+1	0.237	1.44	0.288	0.0209
2	+1	+1	−1	−1	+1	+1	0.160	0.99	0.397	0.0107
3	−1	+1	+1	+1	−1	+1	0.237	1.23	0.289	0.0077
4	−1	−1	+1	+1	−1	−1	0.236	1.45	0.288	0.0082
5	−1	+1	−1	−1	+1	−1	0.178	1.33	0.442	0.0104
6	+1	+1	−1	+1	+1	−1	0.157	0.96	0.395	0.0080
7	−1	+1	−1	+1	+1	+1	0.178	0.57	0.443	0.0071
8	−1	−1	+1	+1	+1	+1	0.178	0.58	0.295	0.0071
9	+1	+1	−1	−1	−1	−1	0.223	1.94	0.386	0.0275
10	+1	−1	+1	−1	−1	−1	0.230	2.03	0.257	0.0274
11	−1	−1	−1	+1	+1	−1	0.178	0.98	0.441	0.0079
12	−1	−1	−1	−1	−1	−1	0.236	1.82	0.432	0.0208
13	−1	−1	+1	−1	+1	−1	0.178	1.33	0.294	0.0105
14	+1	−1	−1	−1	−1	+1	0.224	1.58	0.387	0.0285
15	−1	+1	+1	−1	+1	+1	0.179	0.99	0.295	0.0097
16	−1	+1	+1	+1	+1	−1	0.178	0.98	0.295	0.0079
17	−1	+1	+1	−1	−1	−1	0.236	1.81	0.288	0.0196
18	−1	−1	−1	+1	−1	+1	0.237	1.23	0.433	0.0077
19	−1	+1	−1	−1	−1	+1	0.237	1.43	0.433	0.0209
20	−1	+1	−1	+1	−1	−1	0.236	1.44	0.432	0.0081
21	+1	−1	−1	−1	+1	−1	0.156	1.39	0.395	0.0116
22	+1	+1	−1	+1	−1	+1	0.224	1.52	0.387	0.0286
23	+1	+1	+1	−1	+1	−1	0.157	1.35	0.264	0.0114
24	+1	−1	+1	+1	−1	+1	0.224	1.28	0.258	0.0095
25	+1	+1	+1	+1	−1	−1	0.223	1.50	0.257	0.0097
26	+1	−1	+1	+1	+1	−1	0.156	0.99	0.263	0.0082
27	+1	−1	−1	+1	−1	−1	0.223	1.54	0.385	0.0099
28	−1	−1	−1	−1	+1	+1	0.178	1.00	0.443	0.0097
29	+1	+1	+1	−1	−1	+1	0.224	1.52	0.258	0.0286
30	+1	+1	+1	+1	+1	+1	0.160	0.56	0.265	0.0072
31	+1	−1	−1	+1	+1	+1	0.159	0.58	0.397	0.0073
32	+1	−1	+1	−1	+1	+1	0.159	1.02	0.265	0.0107

reduces the rate of cooling and allows more time for reduction of thermal gradients within the part and gradual part solidification. Also, for semi-crystalline polymer such as HDPE, a reduced rate of cooling improves the molecular alignments of the polymer chains which compensates for volume reduction resulting to a more compact structure [35].

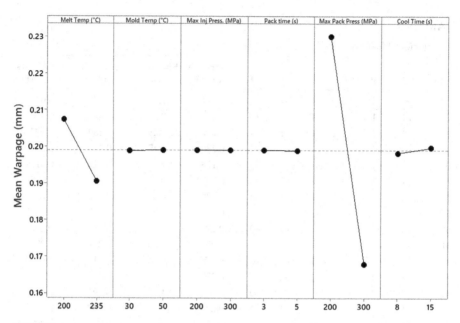

FIGURE 3.1
Mains effect plot for warpage.

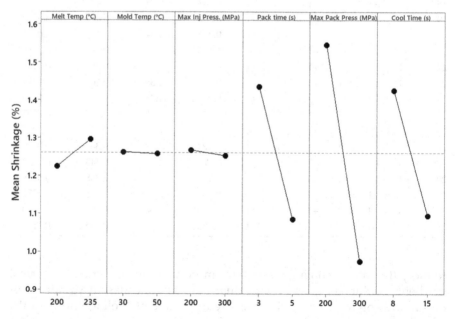

FIGURE 3.2
Shrinkage mains effect plot.

An increase in packing pressure and packing time results to a significant decrease in shrinkage. Increasing packing pressure and time ensures delivery of more polymer melt materials into the mold during packing to compensate for the shrinkage of the material as a result of contact with the colder walls of the mold. Material compensation eliminates voids and gaps within the part which would result to shrinkage on cooling.

An increase in melt temperature increases shrinkage. Higher melt temperatures lowers the material viscosity and increases its fluidity. High material fluidity could thus result to greater shrinkage upon cooling as a result of the higher thermal contraction associated with the elevated temperature. However, an increase in mold temperature slightly reduces the shrinkage in that higher mold temperature slows down the rate of cooling which minimizes thermal gradients and induce controlled solidification. Also, warmer mold surfaces improves material flow and packing thereby minimizing shrinkage.

An increase in injection pressure decreases the shrinkage as a result of enhanced material flow. Pressure drops incurred along the feed system tends to slow the flow of the material into the cavities. Therefore higher injection pressures enhances delivery of the melt to completely fill all the intricate parts of the cavity and improve material packing thus minimizing shrinkage.

3.5.3.3 Short Shot

Mains effect plot of short shot is illustrated on Figure 3.3 and indicate a general decrease in short shot possibility with an increase in melt temperature and injection pressure. A study by Moayyedian et al. [25] obtained similar trends in terms of a decrease in short shot possibility with an increase in melt temperature and slight increase in short shot possibility with increasing cooling time and holding times.

Higher melt temperatures lowers the molten material viscosity making it flow easily into the mold cavity hence reducing the likelihood of incomplete filling and short shots [36]. For complex geometries such the product used in this study, increasing the injection pressures helps to overcome flow resistance brought about by features such as undercuts thereby improving material flowability which lowers chances of incomplete fill and short shot [37].

3.5.3.4 Sink Marks

Figure 3.4 illustrates the mains effect plot for sink mark defect and showing decreasing effect on sink mark depth with increasing packing time, packing pressure and injection pressure whereas an increase in melt temperature, mold temperature and cooling time have an increasing effect on sink mark depth. Parameters with the largest effect on sink mark defect are packing pressure, packing time and melt temperature.

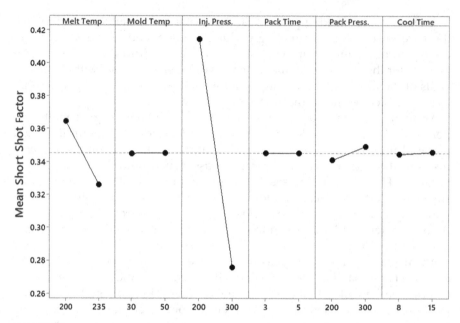

FIGURE 3.3
Mains effect plot for short shot defect.

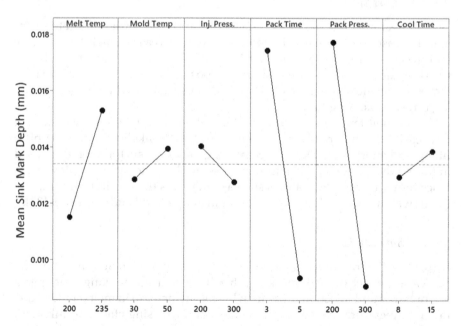

FIGURE 3.4
Mains effect plot for sink mark defect.

Higher packing pressure and packing time ensures the polymer melt is packed uniformly and densely and compensating for material backflow and shrinkage thereby reducing sink mark depth [22]. It also reduces voids and air pockets whose collapse results to sink marks. Higher melt temperature and mold temperature increases internal stresses within the material which causes distortion during cooling thereby increasing the chances of sink mark defects.

3.5.3.5 Effect Sizes

The mains effect sizes were computed based on a difference of means and are shown in Table 3.3. Increasing the melt temperature from 200 to 235°C decreases the warpage by 0.017 mm, increases the shrinkage by 0.071%, decreases short shot possibility factor by 0.038 and increases sink mark depth by 0.004. Increasing the cooling time from 8 to 15 seconds increases the warpage by 0.001 mm and decreases the shrinkage by 0.330%. Also, an increase in the packing pressure from 200 MPa to 300 MPa decreases the warpage by 0.062 mm and decreases the shrinkage by 0.572%. In relation to the small cap specimen under study with dimensions such as a height of 12 mm, minimum wall thickness of 0.03 mm, maximum wall thickness of 2 mm and maximum diameter of 30 mm, these sizes were sufficient.

Although warpage, sink mark and shrinkage defects are interrelated, the responses of the defects are dissimilar for particular process parameter changes. The study establishes that changes in parameters such as melt temperature, mold temperature, injection pressure and cooling time results to a decrease in one defect and an increase in the other. Increasing the cooling time reduces the cooling rate which significantly reduces shrinkage defect but potentially induces warpage due to increased residual stresses.

3.5.4 Interaction Effects

3.5.4.1 Warpage

Figure 3.5 shows the Pareto chart of the mains and interaction effects affecting warpage showing the absolute standardized effect of up to 30 variables

TABLE 3.3

Mains Effect Sizes

Mains Effect	A	B	C	D	E	F
Warpage (mm)	−0.017	1E−04	3E−06	−4E−05	−0.062	0.001
Shrinkage (%)	0.071	−0.005	−0.015	−0.350	−0.572	−0.331
Short shot factor	−0.038	−3E−04	−0.138	−6E−05	0.008	0.001
Sink mark depth (mm)	0.004	0.001	−0.001	−0.008	−0.009	0.001

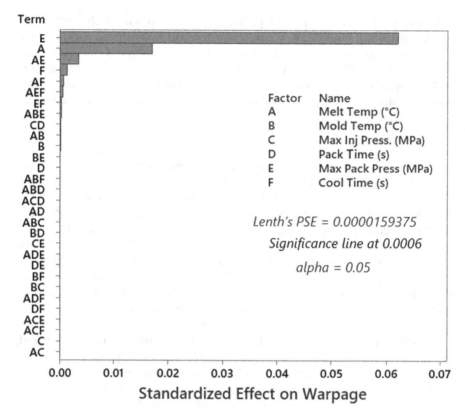

FIGURE 3.5
Pareto chart of the effects of process parameters on warpage.

in the order of significance at a confidence level of 0.05. From the chart, packing pressure is the most significant parameter while the most significant interaction occurs between melt temperature and packing pressure. Injection pressure is the least significant parameter. Also, most three-way interactions do not largely affect warpage.

The chart clearly shows the importance of consideration of the various parameter interaction effects. For instance, the effect of the interaction between injection pressure and packing time on warpage is higher than the individual effects of the same parameters on warpage. Of the 30 effects shown on the chart, the highest number of largest interaction terms involve the melt temperature despite packing pressure being the most significant factor affecting warpage. This could be because melt temperature affect properties of the polymer melt such as flow, viscosity, molecular orientation and thermal behavior. Therefore, an interaction between melt temperature and other process parameters may influence these properties and significantly contribute to warpage.

TABLE 3.4

ANOVA for Warpage Defect

Source	DF	Seq SS	Contribution (%)	Adj MS	F-Value	P-Value
Melt Temp	1	2.3E-03	6.95	2.3E-03	8243.15	0.000
Mold Temp	1	1.9E-07	0.00	1.9E-07	0.66	0.426
Inj. Press.	1	7.8E-11	0.00	7.8E-11	0.00	0.987
Pack Time	1	1.9E-08	0.00	1.9E-08	0.07	0.794
Pack Press.	1	3.1E-02	92.68	3.1E-02	109952.48	0.000
Cool Time	1	1.6E-05	0.05	1.6E-05	58.33	0.000
Melt Temp*Pack Press.	1	9.8E-05	0.30	9.8E-05	351.74	0.000
Melt Temp*Cool Time	1	3.7E-06	0.01	3.7E-06	13.20	0.002
Inj. Press.*Pack Press.	1	1.4E-09	0.00	1.4E-09	0.00	0.945
Pack Time*Cool Time	1	1.5E-10	0.00	1.5E-10	0.00	0.982
Error	21	5.9E-06	0.02	2.8E-07		
Total	31	3.3E-02	100.00			

Table 3.4 shows the ANOVA results of warpage. To verify the contribution of interaction terms to warpage, ANOVA was carried out first with main terms only, then with interaction terms. ANOVA with main terms only yielded R-squared of 99.67%, adjusted R-squared of 99.60% and predicted R-squared of 99.47% whereas ANOVA with interaction terms yielded R-squared of 99.98%, adjusted R-squared of 99.96% and predicted R-squared of 99.91%. The increase in adjusted R-squared with the addition of interaction terms indicated that they improved the model and the smaller difference between R-squared and predicted R-squared implied minimal chances of model overfitting. For warpage defect, the most significant parameters were packing pressure, melt temperature and cooling time while the most significant interactions involved the three parameters.

Table 3.5 shows the significant interaction sizes affecting warpage computed. A positive effect indicates an additional increasing effect on warpage while a negative effect indicates an additional decreasing effect on warpage.

TABLE 3.5

Interaction Effect Sizes on Warpage

Parameter Interaction	Effect on Warpage (mm)
Melt Temp × Pack Press	−0.004
Melt Temp × Cool Time	0.0007
Melt Temp × Pack Press × Cool Time	0.0007
Pack Press × Cool Time	0.0004
Melt Temp × Mold Temp × Pack Press	0.0003

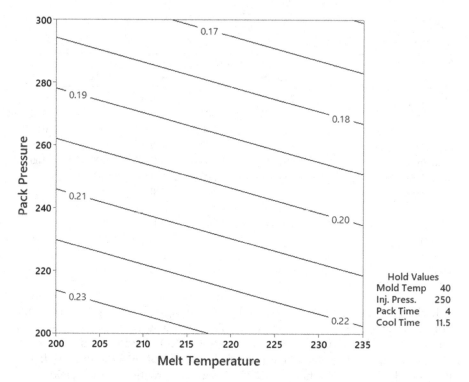

FIGURE 3.6
Effect of packing pressure and melt temperature on warpage.

An interaction between melt temperature and packing pressure reduces warpage further by 0.004 mm. This effect is better shown by contour plot in Figure 3.6. From the contour plot, the highest values of warpage are obtained at lower values of packing pressure and melt temperatures while the lowest warpage obtained at higher values of the two parameters.

Also, the effect of variation of one parameter to warpage depends on the level of the other parameter. At lower packing pressure, an increase in melt temperature has no effect on warpage whereas at higher packing pressure, an increase in melt temperature substantially lowers the warpage. Conversely, increasing the packing pressure decreases warpage both at lower and higher levels of melt temperature with the largest decrease at higher melt temperature. This is confirmed by the size computation where the mains effect of packing pressure on warpage at a melt temperature of 200°C was found to be −0.059 mm while at a melt temperature of 235°C it became −0.066 mm. Higher melt temperature lowers the material viscosity and increases its fluidity thereby enhancing easier molecular orientation and reduction in internal stresses at higher packing pressure. Therefore, when all the other factors are held constant, warpage defect would be reduced significantly by raising the packing pressure at higher melt temperatures.

A combined effect of melt temperature and cooling time increases warpage by 0.0007 mm. In plastic injection molding environment, the cooling time affects the rate of cooling which in turn depends on the melt temperature. Increasing the melt temperature at higher level of cooling time lowers warpage more than at lower level of cooling time.

3.5.4.2 Shrinkage

Pareto chart of the effect of process parameters affecting shrinkage is shown in Figure 3.7. Packing pressure has the largest effect followed by packing time. The largest interaction is a three-way interaction between melt temperature, mold temperature and injection pressure.

The effect of interaction between melt temperature, mold temperature and injection pressure on shrinkage is larger than the individual effects of the three parameters. Also, many interactions of larger size involve the melt

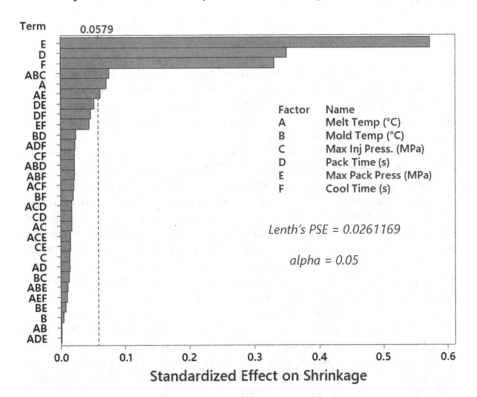

Standardized Effect on Shrinkage

FIGURE 3.7
Pareto chart of the effects of process parameters on shrinkage.

TABLE 3.6

ANOVA for Shrinkage Defect

Source	DF	Seq SS	Contrib. (%)	Adj MS	F-Value	P-Value
Melt Temp	1	4.0E-02	0.86	4.0E-02	12.88	0.002
Mold Temp	1	2.2E-04	0.00	2.2E-04	0.07	0.795
Inj. Press.	1	1.7E-03	0.04	1.7E-03	0.55	0.466
Pack Time	1	9.8E-01	20.87	9.8E-01	313.40	0.000
Pack Press.	1	2.6E+00	55.83	2.6E+00	838.62	0.000
Cool Time	1	8.8E-01	18.67	8.8E-01	280.45	0.000
Melt Temp*Mold Temp	1	2.9E-05	0.00	2.9E-05	0.01	0.925
Melt Temp*Inj. Press.	1	2.4E-03	0.05	2.4E-03	0.77	0.392
Melt Temp*Pack Press.	1	3.0E-02	0.65	3.0E-02	9.76	0.006
Mold Temp*Inj. Press.	1	1.5E-03	0.03	1.5E-03	0.47	0.503
Pack Time*Pack Press.	1	2.2E-02	0.47	2.2E-02	7.09	0.016
Pack Press.*Cool Time	1	1.6E-02	0.34	1.6E-02	5.10	0.037
Melt Temp*Mold Temp*Inj. Press.	1	4.6E-02	0.99	4.6E-02	14.81	0.001
Error	18	5.6E-02	1.20	3.1E-03		
Total	31	4.7E+00	100.00			

temperature despite packing pressure being the most significant parameter affecting shrinkage. This is because melt temperature affects the material viscosity, flow and thermal behavior and therefore an interaction between the other process parameters and melt temperature would largely affect these properties and influence shrinkage.

Table 3.6 shows ANOVA results of shrinkage. To verify the contribution of interaction terms to shrinkage, ANOVA was carried out initially with main terms only, then with interaction terms. ANOVA with main terms only yielded R-squared of 96.27%, adjusted R-squared of 95.38% and predicted R-squared of 93.89% whereas ANOVA with interaction terms yielded R-squared of 98.8%, adjusted R-squared of 97.94% and predicted R-squared of 96.21%. Similar to warpage, the increase in adjusted R-squared with the addition of interaction terms indicated that they improved the model and the smaller difference between R-squared and predicted R-squared implied minimal chances of model overfitting.

A significant interaction occurred between melt temperature and other less significant parameters such as mold temperature and injection pressure. Interaction effect sizes on shrinkage between the three parameters were found to be −0.07, that between melt temperature and injection pressure −0.02, between mold temperature and injection pressure −0.01 and between melt temperature and mold temperature found to be −0.002.

Table 3.7 shows some of the major interaction sizes for shrinkage computed. The combined effect of increasing the melt temperature, mold temperature

TABLE 3.7

Interaction Effect Sizes on Shrinkage

Parameter Interaction	Effect on Shrinkage (%)
Melt Temp × Mold Temp × Inj. Press	−0.0760
Melt Temp × Pack Press	−0.0620
Pack Time × Pack Press	−0.0526
Pack Time × Cool Time	0.0467
Pack Press × Cool Time	−0.0450
Mold Temp × Pack Time	0.0233
Melt Temp × Pack Time × Cool Time	0.0228

and injection pressure is the most significant interaction and reduces shrinkage further by 0.076%. This is a double interaction between melt temperature and an interaction between mold temperature and injection pressure. The interaction between these factors affect the material flow into the cavity. Higher melt temperature as well as mold temperature enhances smooth material flow into the cavities. A higher injection pressure then ensures delivery of the smoothly flowing molten material to all sections of the cavity thus reducing shrinkage.

A similar relationship was obtained between melt temperature and packing pressure whose interaction equally have a significant contribution to shrinkage. Individually, increasing melt temperature increases shrinkage while increasing packing pressure reduces shrinkage. When the two parameters are increased simultaneously, the shrinkage reduces but the extent of reduction depends on the levels of the parameters. At a melt temperature of 200°C, an increase in packing pressure from 200 to 300 MPa would decrease shrinkage by 0.633% while at a melt temperature of 235°C, the same increase in packing pressure would decrease shrinkage by 0.510%.

A combined effect of packing time and packing pressure on shrinkage is also imminent as a result of the contribution of packing phase to shrinkage. Figure 3.8 shows a contour plot of packing pressure and time against shrinkage. As shown on the contour plot, minimum shrinkage rate is obtained at higher packing pressure and packing time. At low level of packing pressure, an increase in packing time decreases shrinkage by 0.3% whereas at high level of packing pressure, the same increase in packing time decreases shrinkage by 0.4%.

3.5.4.3 Short Shot

Figure 3.9 illustrates a Pareto chart of standardized effects on short shot defect. Injection pressure setting has the largest effect on short shot possibility followed by melt temperature while the largest interaction effect involved

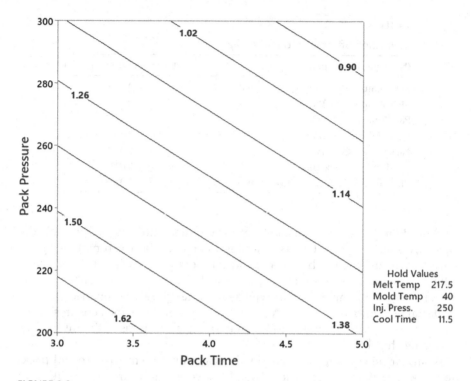

FIGURE 3.8
Effect of packing pressure and packing time on shrinkage.

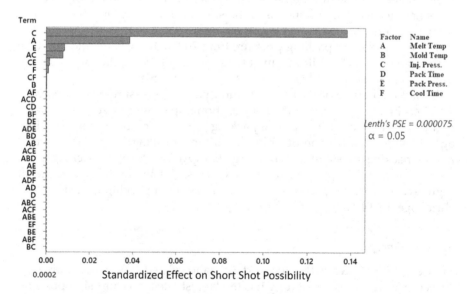

FIGURE 3.9
Pareto chart of standardized effects on short shot possibility.

TABLE 3.8

ANOVA for Short Shot Defect

Source	DF	Adj SS	Adj MS	F-Value	P-Value
Melt Temp	1	1.2E-02	1.2E-02	553323.7	0.000
Mold Temp	1	6.6E-07	6.6E-07	30.82	0.000
Inj. Press.	1	1.5E-01	1.5E-01	7116630	0.000
Pack Time	1	2.0E-08	2.0E-08	0.93	0.347
Pack Press.	1	5.5E-04	5.5E-04	25528.78	0.000
Cool Time	1	1.4E-05	1.4E-05	654.52	0.000
Melt Temp*Inj. Press.	1	4.8E-04	4.8E-04	22464.52	0.000
Melt Temp*Pack Time	1	2.0E-08	2.0E-08	0.93	0.347
Melt Temp*Cool Time	1	1.8E-07	1.8E-07	8.39	0.010
Inj. Press.*Pack Time	1	6.1E-08	6.1E-08	2.85	0.108
Inj. Press.*Pack Press.	1	2.0E-05	2.0E-05	939.55	0.000
Inj. Press.*Cool Time	1	9.1E-07	9.1E-07	42.47	0.000
Melt Temp*Inj. Press.*Pack Time	1	1.0E-07	1.0E-07	4.72	0.043
Error	18	3.9E-07	2.2E-08		
Total	31	1.7E-01			

the two parameters. The largest three way interaction involved the melt temperature, injection pressure and packing time. Individual effect of a parameter like packing time is lower than the effects of most interactions involving the parameter.

Table 3.8 shows ANOVA results for short shot defect. Only packing time does not significantly affect short shot factor. However, some interactions involving packing time are significant.

Some of the major computed interaction effects were −0.008 for melt temperature by injection pressure, −0.002 for injection pressure by packing pressure, −0.0003 for injection pressure by cooling time and −0.0001 for melt temperature by injection pressure by packing time. A combined effect of increasing the melt temperature and injection pressure simultaneously reduces short shot possibility ratio further by 0.008. Melt temperature influences material viscosity while injection pressure influences material flow. At higher melt temperature and injection pressure, material delivery into the mold cavity is enhanced and thus less short shot possibility. Therefore, a well-balanced and jointly optimized melt temperature and injection pressure would help overcome short shot defects.

The interaction between melt temperature and injection pressure is illustrated on a contour plot on Figure 3.10. Short shot possibility is higher at lower melt temperature and injection pressure and lower at higher values.

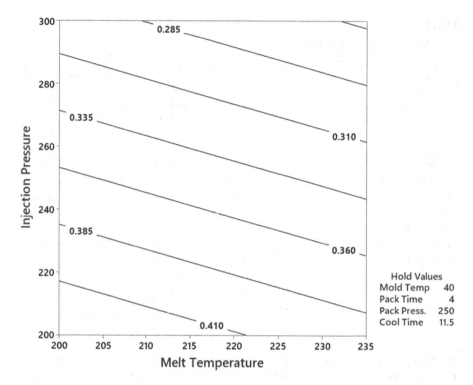

FIGURE 3.10
Contour plot of melt temperature against injection pressure setting.

Contour plot trends are not horizontal indicating a significant interaction effect between the two parameters. At constant melt temperature, an increase in injection pressure reduces short shot possibility but the rate of reduction depends on the level of melt temperature. At 200°C, increasing injection pressure from 200 MPa to 250 MPa reduces short shot possibility factor to 0.37 whereas at 235°C, a similar increase in injection pressure reduces short shot possibility factor to 0.33. With such information about parameters with the largest interaction effect on short shot, a targeted approach to the defect control could be used.

3.5.4.4 Sink Mark

Pareto chart on Figure 3.11 illustrates effect of process parameter interactions on sink mark depth. The largest two way interaction involve packing pressure and packing time whereas the largest three way interaction involve melt temperature, mold temperature and packing pressure. The effect of interaction between mold temperature and cooling time on sink mark is larger than the individual effects of each of the two parameters on sink mark.

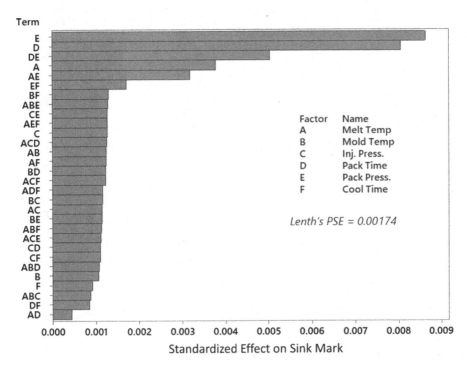

FIGURE 3.11
Pareto chart of the effects of process parameters on sink mark.

Figure 3.12 illustrates the major interaction effects affecting sink mark defect. Larger interaction effect occurs between melt temperature, packing time and packing pressure whereas the interaction between melt temperature and packing time is minimal. The effect of variation of packing time depends on the level of packing pressure. At 200 MPa packing pressure, an increase in packing time from 3s to 5s reduces sink mark depth by 0.016 whereas a similar increase in packing time reduces sink mark depth by 0.008 at 300 MPa packing pressure. This implies that the effect of variation of packing time on sink mark defect is maximized at lower levels of packing pressure and minimized at higher levels. Similarly, the effect of variation of melt temperature is maximized at lower levels of packing pressure and minimized at higher levels. Major interaction sizes computed included 0.005 for packing pressure by packing time and −0.003 for packing pressure by melt temperature.

Table 3.9 shows ANOVA results for sink mark defect. ANOVA with main terms only yielded R-squared of 70.33%%, adjusted R-squared of 63.21% and predicted R-squared of 51.40% whereas ANOVA with interaction terms yielded R-squared of 86.27%, adjusted R-squared of 81.50% and predicted R-squared of 73.43%. Similar to the other defect, the increase in adjusted R-squared with the addition of interaction terms indicated that they

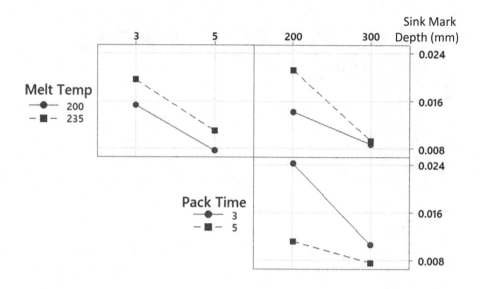

FIGURE 3.12
Major interaction effects affecting sink mark.

TABLE 3.9

ANOVA for Sink Mark Defect

Source	DF	Seq SS	% Cont. (%)	Adj SS	Adj MS	F-Value	P-Value
Melt Temp	1	0.000115	6.41	0.000115	0.000115	10.74	0.003
Mold Temp	1	0.000009	0.52	0.000009	0.000009	0.87	0.362
Inj. Press.	1	0.000013	0.73	0.000013	0.000013	1.22	0.281
Pack Time	1	0.000518	28.96	0.000518	0.000518	48.52	0.000
Pack Press.	1	0.000597	33.34	0.000597	0.000597	55.86	0.000
Cool Time	1	0.000007	0.38	0.000007	0.000007	0.64	0.432
Melt Temp*Pack Press.	1	0.000081	4.54	0.000081	0.000081	7.61	0.011
Pack Time*Pack Press.	1	0.000204	11.40	0.000204	0.000204	19.09	0.000
Error	23	0.000246	13.73	0.000246	0.000011		
Total	31	0.001790	100.00				

improved the model and the smaller difference between R-squared and predicted R-squared implied minimal chances of model overfitting.

3.5.5 Effects on Molding Defect Index

As plastic injection molded products may be subject to a variety of defects occurring simultaneously during the molding period, it may be necessary on some occasions to create a single composite index by weighting multiple defects. Injection molded products are rarely affected by a single defect as they are oftenly subjected to multiple defects thus considering these defects collectively could offer a more comprehensive assessment. Investigation of the effects of variations in process parameters to the weighted index would allow for a holistic evaluation.

Considered in the case study were four of the major defects encountered in plastic injection molded products. These included warpage, shrinkage, short shot and sink marks. A single composite defect index was computed based on the four defects. Relative importance of each of the four defects were assessed and weights assigned based on expert opinion. With respect to a packaging bottle cap specimen considered in the study, design for assembly guidelines requires strict conformity of the product to the design tolerances, otherwise a tight package seal would not be achieved.

A short shot defect could result to an incomplete formed cap which could render the cap unusable hence allocated the highest weighting of 35%. Warpage and shrinkage defects directly impacts on the shape and dimensions of the cap potentially affecting its ability to form a tight seal hence both allocated weightings of 25% each. Sink mark defects have a potential effect on the cap surface integrity and aesthetics whose effect on the cap functionality is lower hence allocated a weighting of 15%.

To ensure all the defects were on the same scale for comparison, all the defect data were normalized on scales from 0 to 1 where 0 indicated the lowest (equivalent to none) while 1 indicated the highest (most severe) defect value. Normalization was based on minimum and maximum values of each of the defects and the normalized value computed based on Equation (3.3). For shrinkage values ranging between 0.56% and 2.03%, a shrinkage value of 1.52% had a normalized value of 0.65.

$$\text{Normalized Value} = \frac{\text{Defect Value} - \text{Minimum Value}}{\text{Maximum Value} - \text{Minimum Value}} \qquad (3.3)$$

A molding defect index was thus computed using the weighted sum of the normalized defects based on Equation (3.4) where W represents the weights in terms of ratios and N represents the normalized values of the defects.

Based on numerical results on Table 3.2, run 12 yielded the highest defect index of 0.886 while run 30 yielded the lowest defect index of 0.028.

$$\text{Index} = W_{\text{warp}} \cdot N_{\text{warp}} + W_{\text{shrink}} \cdot N_{\text{shrink}} + W_{\text{sink}} \cdot N_{\text{sink}} + W_{\text{short}} \cdot N_{\text{short}} \qquad (3.4)$$

Mains effect plot illustrated on Figure 3.13 indicates a decrease in molding defect index with an increase in melt temperature, injection pressure, packing time, packing pressure and cooling time. This implies that the four defects could be simultaneously lowered at higher values of the five process parameters and lower values of mold temperature.

Figure 3.14 shows Pareto plot of standardized effects of process parameters to the molding defect index. Packing pressure has the largest effect on the defect index followed closely by injection pressure indicating that packing pressure and injection pressure have the largest effect on all the four defects considered in this study. A two-way interaction with the largest influence on the molding index involved the melt temperature and packing pressure whereas a three-way interaction with the highest influence involved melt temperature, mold temperature and injection pressure.

ANOVA results for the molding defect index illustrated on Table 3.10 indicated that only the mold temperature had no significant impact on the changes in molding defect index whereas interactions involving melt temperature, packing pressure and packing time significantly contributed to the changes in molding defect index.

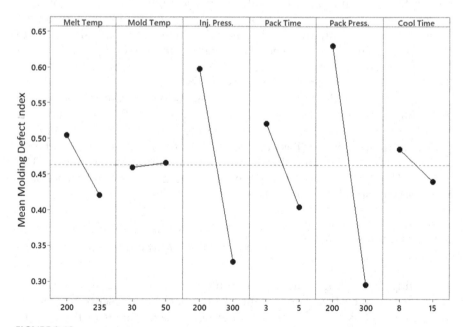

FIGURE 3.13
Molding index mains effect plot.

Term

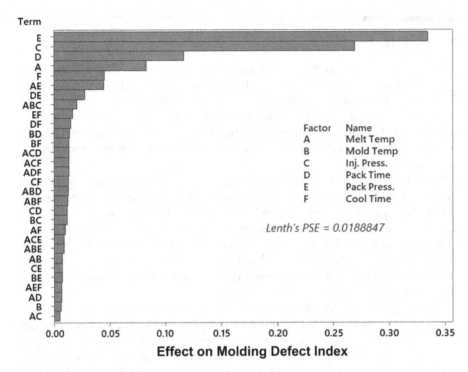

Effect on Molding Defect Index

FIGURE 3.14
Pareto chart of the effect of process parameters on defect index.

TABLE 3.10

ANOVA for Molding Defect Index

Source	DF	Seq SS	Contribution (%)	Adj SS	Adj MS	F-Value	P-Value
Melt Temp	1	0.05585	3.27	0.055852	0.055852	50.03	0.000
Mold Temp	1	0.00033	0.02	0.000330	0.000330	0.30	0.592
Inj. Press.	1	0.58190	34.06	0.581897	0.581897	521.19	0.000
Pack Time	1	0.10976	6.42	0.109765	0.109765	98.31	0.000
Pack Press.	1	0.89638	52.47	0.896377	0.896377	802.86	0.000
Cool Time	1	0.01631	0.95	0.016315	0.016315	14.61	0.001
Melt Temp*Pack Press.	1	0.01612	0.94	0.016116	0.016116	14.43	0.001
Pack Time*Pack Press.	1	0.00607	0.36	0.006075	0.006075	5.44	0.029
Error	23	0.02568	1.50	0.025679	0.001116		
Total	31	1.70841	100.00				

3.6 Conclusion

The advancements in plastic injection molding have been significantly compelled by the application of statistical modelling tools and techniques. Utilization of various design of experiment methodologies has brought about systematic approaches to understanding and optimizing various process parameters and variables affecting product quality. This has enabled manufacturers to produce higher quality, defect-free products for maximum returns.

Response surface methodology has provided structured and systematic approach to the analysis of the relationship between process parameters and the molded product quality. Taguchi method emphasizes robustness against variations and reductions of sensitivity to external factors. Factorial design offers a comprehensive approach to studying mains effects and interaction effects among process parameters and allows manufacturers to examine the impacts of each parameter and their interactions which enhances informed decisions about ideal parameter settings.

Studying interactions effects alongside mains effects is vital in establishing how combined effects of parameters impacts the final product. Through comprehensive evaluation of the effects, manufacturers obtain more insights into the manufacturing process and enables them to optimize the settings for quality product with minimized defects. Through interaction effect sizes, a targeted approach to the defects minimization can be applied.

Therefore, knowledge of effects of process parameters to major plastic injection molding defects such as warpage, shrinkage, sink marks and short shots is crucial for achieving high quality products. Warpage and shrinkage can be reduced by adjusting major parameters packing pressure, melt temperature and cooling time whereas sink marks and short shots are better reduced by maintaining the right melt temperature, packing pressure and injection pressure. Parameters like melt temperature and packing pressure significantly contribute to all the four defects indicating their crucial importance to the plastic injection molding process. The quest to understand and optimize process parameter settings to control these (and other related) defects remains the central theme in plastic injection molding research.

References

[1] C. Fernandes, A. J. Pontes, J. C. Viana, and A. Gaspar-Cunha, "Modeling and optimization of the injection-molding process: A review," *Adv. Polym. Technol.*, vol. 37, no. 2, pp. 429–449, 2018. doi: 10.1002/adv.21683

[2] S. M. S. Mukras, H. M. Omar, and F. A. Al-Mufadi, "Experimental-based multi-objective optimization of injection molding process parameters," *Arab. J. Sci. Eng.*, vol. 44, no. 9, pp. 7653–7665, 2019. doi: 10.1007/s13369-019-03855-1

[3] M. H. N. Hidayah, Z. Shayfull, N. Z. Noriman, M. Fathullah, R. Norshahira, and A. T. N. A. Miza, "Optimization of warpage on plastic part by using genetic algorithm (GA)," *AIP Conf. Proc.*, vol. 2030, no. November, 2018. doi: 10.1063/1.5066804

[4] D. Mathivanan and N. S. Parthasarathy, "Prediction of sink depths using non-linear modeling of injection molding variables," *Int. J. Adv. Manuf. Technol.*, 2009. doi: 10.1007/s00170-008-1749-1

[5] W. C. Lin, F. Y. Fan, C. F. Huang, Y. K. Shen, and H. Wang, "Analysis of the warpage phenomenon of micro-sized parts with precision injection molding by experiment, numerical simulation, and grey theory," *Polymers (Basel)*, vol. 14, no. 9, 2022. doi: 10.3390/polym14091845

[6] G. Trotta, S. Cacace, and Q. Semeraro, "Optimizing process parameters in micro injection molding considering the part weight and probability of flash formation," *J. Manuf. Process.*, vol. 79, pp. 250–258, Jul. 2022. doi: 10.1016/j.jmapro.2022.04.048

[7] K. Rajendra, H. Vasudevan, and G. Vimal, "Optimization of injection molding process parameters using response surface methodology," *Lect. Notes Mech. Eng.*, pp. 445–454, 2019. doi: 10.1007/978-981-13-2490-1_40/COVER

[8] D. Mathivanan and N. S. Parthasarathy, "Sink-mark minimization in injection molding through response surface regression modeling and genetic algorithm," *Int. J. Adv. Manuf. Technol.*, vol. 45, no. 9–10, pp. 867–874, Dec. 2009. doi: 10.1007/s00170-009-2021-z

[9] R. Azad and H. Shahrajabian, "Experimental study of warpage and shrink-age in injection molding of HDPE/rPET/wood composites with multiobjec-tive optimization," *Mater. Manuf. Process.*, vol. 34, no. 3, pp. 274–282, 2019. doi: 10.1080/10426914.2018.1512123

[10] U. M. R. Paturi, S. K. P. Kolluru, and S. D. S. A. Kalvakolanu, "Prediction of weld-line width and sink-mark depth of plastic injection moulded parts using neural networks," *Mater. Today Proc.*, Mar. 2023. doi: 10.1016/j.matpr.2023.02.295

[11] X. Liu, X. Fan, Y. Guo, B. Man, and L. Li, "Warpage optimization of the GFRP injection molding process parameters," *Microsyst. Technol.*, vol. 27, no. 12, pp. 4337–4346, Dec. 2021. doi: 10.1007/s00542-021-05241-0

[12] D. Wang, X. Fan, Y. Guo, X. Lu, C. Wang, and W. Ding, "Predicting and con-trolling the quality of injection molding properties for fiber-reinforced com-posites," *SAE Int. J. Mater. Manuf.*, vol. 16, no. 3, pp. 293–306, Apr. 2023. doi: 10.4271/05-16-03-0020

[13] Y. K. Yang, J. R. Shie, H. Te Liao, J. L. Wen, and R. T. Yang, "A study of Taguchi and design of experiments method in injection molding process for polypropyl-ene components," *J. Reinf. Plast. Compos.*, vol. 27, no. 8, pp. 819–834, May 2008. doi: 10.1177/0731684407084988

[14] N. Y. Zhao, J. Y. Lian, P. F. Wang, and Z. Bin Xu, "Recent progress in minimiz-ing the warpage and shrinkage deformations by the optimization of process parameters in plastic injection molding: A review," *Int. J. Adv. Manuf. Tech.*, vol. 120, no. 1–2, pp. 85–101, Feb. 10, 2022. doi: 10.1007/s00170-022-08859-0

[15] E. Hakimian and A. B. Sulong, "Analysis of warpage and shrinkage properties of injection-molded micro gears polymer composites using numerical simulations assisted by the Taguchi method," *Mater. Des.*, vol. 42, pp. 62–71, Dec. 2012. doi: 10.1016/J.MATDES.2012.04.058

[16] K. Li, S. Yan, Y. Zhong, W. Pan, and G. Zhao, "Multi-objective optimization of the fiber-reinforced composite injection molding process using Taguchi method, RSM, and NSGA-II," *Simul. Model. Pract. Theory*, vol. 91, pp. 69–82, Feb. 2019. doi: 10.1016/J.SIMPAT.2018.09.003

[17] M. Altan, "Reducing shrinkage in injection moldings via the Taguchi, ANOVA and neural network methods," *Mater. Des.*, vol. 31, no. 1, pp. 599–604, Jan. 2010. doi: 10.1016/J.MATDES.2009.06.049

[18] G. Singh, M. K. Pradhan, and A. Verma, "Multi response optimization of injection molding process parameters to reduce cycle time and warpage," *Mat. Today: Proc.*, pp. 8398–8405, Jan. 2018. doi: 10.1016/j.matpr.2017.11.534

[19] E. Oliaei *et al.*, "Warpage and shrinkage optimization of injection-molded plastic spoon parts for biodegradable polymers using Taguchi, ANOVA and artificial neural network methods," *J. Mater. Sci. Technol.*, vol. 32, no. 8, pp. 710–720, Aug. 2016. doi: 10.1016/j.jmst.2016.05.010

[20] P. Pachorkar, G. Singh, N. Agarwal, and A. Srivastava, "Multi response optimization of injection molding process to reduce sink marks and cycle time," *Mater. Today Proc.*, vol. 72, pp. 1089–1093, Jan. 2023. doi: 10.1016/j.matpr.2022.09.172

[21] D. Mathivanan, M. Nouby, and R. Vidhya, "Minimization of sink mark defects in injection molding process – Taguchi approach," *Int. J. Eng. Sci. Technol.*, vol. 2, no. 2, pp. 13–22, Sep. 2010. doi: 10.4314/ijest.v2i2.59133

[22] C. Shen, L. Wang, W. Cao, and L. Qian, "Investigation of the effect of molding variables on sink marks of plastic injection molded parts using Taguchi DOE technique," *Polym. – Plast. Technol. Eng.*, vol. 46, no. 3, pp. 219–225, Mar. 2007. doi: 10.1080/03602550601152887

[23] W. Guo, L. Hua, and H. Mao, "Minimization of sink mark depth in injection-molded thermoplastic through design of experiments and genetic algorithm," *Int. J. Adv. Manuf. Technol.*, vol. 72, no. 1–4, pp. 365–375, Feb. 2014. doi: 10.1007/s00170-013-5603-8

[24] J. Sreedharan and A. K. Jeevanantham, "Optimization of injection molding process to minimize weld-line and sink-mark defects using Taguchi based grey relational analysis," in *Mat. Today: Proc.*, pp. 12615–12622, Jan. 2018. doi: 10.1016/j.matpr.2018.02.244

[25] M. Moayyedian, K. Abhary, and R. Marian, "The analysis of short shot possibility in injection molding process," *Int. J. Adv. Manuf. Technol.*, vol. 91, no. 9–12, pp. 3977–3989, Feb. 2017. doi: 10.1007/s00170-017-0055-1

[26] E. A. Wibowo, A. Syahriar, and A. Kaswadi, "Analysis and simulation of short shot defects in plastic injection molding at multi cavities," in *ACM International Conference Proceeding Series*, Association for Computing Machinery, Sep. 2020. doi: 10.1145/3429789.3429837

[27] J. Heinisch, Y. Lockner, and C. Hopmann, "Comparison of design of experiment methods for modeling injection molding experiments using artificial neural networks," *J. Manuf. Process.*, vol. 61, pp. 357–368, Jan. 2021. doi: 10.1016/J.JMAPRO.2020.11.011

[28] M. Packianather, F. Chan, C. Griffiths, S. Dimov, and D. T. Pham, "Optimisation of micro injection molding process through design of experiments," *Procedia CIRP*, vol. 12, pp. 300–305, Jan. 2013. doi: 10.1016/J.PROCIR.2013.09.052

[29] M. A. Barghash and F. A. Alkaabneh, "Shrinkage and warpage detailed analysis and optimization for the injection molding process using multistage experimental design," *Qual. Eng.*, vol. 26, no. 3, pp. 319–334, Jul. 2014. doi: 10.1080/08982112.2013.852679

[30] A. Kramschuster, R. Cavitt, D. Ermer, Z. B. Chen, and L. S. Turng, "Effect of processing conditions on shrinkage and warpage and morphology of injection moulded parts using microcellular injection molding," *Plast. Rubber Compos.*, vol. 35, no. 5, pp. 198–209, 2006. doi: 10.1179/174328906X128199

[31] F. K. H. Phoa, W. K. Wong, and H. Xu, "The need of considering the interactions in the analysis of screening designs," *J. Chemom.*, vol. 23, no. 10, pp. 545–553, Oct. 2009. doi: 10.1002/CEM.1252

[32] J. M. Wakiru, L. Pintelon, P. N. Muchiri, and P. K. Chemweno, "A simulation-based optimization approach evaluating maintenance and spare parts demand interaction effects," *Int. J. Prod. Econ.*, vol. 208, no. 2013, pp. 329–342, 2019. doi: 10.1016/j.ijpe.2018.12.014

[33] Y. Chen and J. Zhu, "Warpage analysis and optimization of thin-walled injection molding parts based on numerical simulation and orthogonal experiment," *IOP Conf. Ser. Mater. Sci. Eng.*, vol. 688, no. 3, 2019. doi: 10.1088/1757-899X/688/3/033027

[34] G. Singh, M. K. Pradhan, and A. Verma, "Multi response optimization of injection molding process parameters to reduce cycle time and warpage," *Mater. Today Proc.*, vol. 5, no. 2, pp. 8398–8405, 2018. doi: 10.1016/j.matpr.2017.11.534

[35] D. Annicchiarico and J. R. Alcock, "Review of factors that affect shrinkage of molded part in injection molding," *Mater. Manuf. Process.*, vol. 29, no. 6, pp. 662–682, 2014. doi: 10.1080/10426914.2014.880467

[36] M. Mohan, M. N. M. Ansari, and R. A. Shanks, "Review on the effects of process parameters on strength, shrinkage, and warpage of injection molding plastic component," *Polym. – Plast. Technol. Eng.*, vol. 56, no. 1, pp. 1–12, 2017. doi: 10.1080/03602559.2015.1132466

[37] S. Kitayama, M. Yokoyama, M. Takano, and S. Aiba, "Multi-objective optimization of variable packing pressure profile and process parameters in plastic injection molding for minimizing warpage and cycle time," *Int. J. Adv. Manuf. Tech.*, vol. 92, no. 9–12, pp. 3991–3999, Oct. 2017. doi: 10.1007/s00170-017-0456-1

4

Predictive Modelling of Injection Molding Defects

4.1 Introduction

The multivariability and non-linearity aspects of plastic injection molding process calls for the application of intelligent approximation algorithms to model the process other than physical models. This has therefore led to application of intelligent modelling algorithms to model and predict various quality indices. Common intelligent approximation algorithms used in defects modelling include neural networks, genetic algorithm, Kriging model and response surface methodology [1].

ANN emulates biological neural systems behavior and have three major layers, the input layer, hidden layer and output layer. A trained neural network model is obtained through supervised learning where a series of input and their corresponding outputs are examined. ANN is one of the most popular warpage and shrinkage prediction models in plastic injection molding owing to its higher predictive precision and accuracy [1].

4.2 Injection Molding Defects Modelling

Complexity, variability and non-linear nature of plastic injection molding process has resulted to the application of intelligent approximation algorithms to model the process other than physical models. This has therefore led to application of intelligent modelling algorithms to model and predict various quality indices. Common intelligent approximation algorithms used in major injection molded product defects modelling include neural networks, genetic algorithm, Kriging model and response surface methodology [1]. ANN emulates biological neural systems behavior and have three major layers, the input layer, hidden layer and output layer. ANN is one of the most

72 DOI: 10.1201/9781003492498-4

popular warpage and shrinkage prediction models in plastic injection molding owing to its higher predictive precision and accuracy [1].

A numerical simulation study by Song et al. [2] utilized Support Vector Machine (SVM) and Genetic Algorithm (GA) model for predictive modelling and optimization of the process parameters for warpage and shrinkage control. The authors utilized response surface method to design an experiment using process parameters such as melt temperature, mold temperature, injection time, holding time, holding pressure and cooling time and carried out simulations to obtain warpage and volumetric shrinkage responses. SVM combined with Genetic Algorithm – Back Propagation (BP-GA) was used to build a prediction model for warpage and volumetric shrinkage. The authors compared predictive capabilities of Neural Networks, BP-GA and SVM-BP-GA thereby establishing that SVM-BP-GA model predictive capability was high with higher accuracy.

Paturi et al. [3] developed a weld-line width and sink mark depth predictive model based on artificial neural networks. The study utilized experimental data specific to polymethyl methacrylate material to train the model based on select injection molding process parameters and inferred that artificial neural networks have a precise predictive capability for weldline and sink mark defects.

A study by Abdul et al. [4] combined Taguchi approach with ANN model to investigate the effects of process parameters such as injection speed, cooling time and holding to length and width shrinkage of a high density polyethylene (HDPE) injection molding. Taguchi experiment design was used to obtain the various combinations of input variables and experiments conducted to determine the length and width shrinkages at the various parameter settings. The study designed, trained, tested and validated an ANN model to predict the various shrinkages and inferred that the ANN model is an effective prediction tool for shrinkage.

Other studies have employed the use of Kriging surrogate model which is based on variation function theory and structural analysis to carry out optimization of regionalized variables. Wang et al. [5] carried out warpage optimization using a Kriging model which was used to formulate a functional relationship between the maximum warpage objective and selected 12 process parameters. Moreover, Gao and Wang [6] used Kriging model to optimize process parameters to reduce warpage in plastic injection molded product. The study designed an experiment and used CAE simulations to analyze warpage at the various process parameter settings. The result from the study showed a significant reduction in warpage with the utilization of Kriging model algorithm.

Table 4.1 highlights a summary of the warpage and shrinkage defects predictive modelling research.

TABLE 4.1

Summary of predictive modelling research

Author	Predicted Outputs	Predictive Model
Hakimian and Sulong [7]	Shrinkage Warpage	Regression
Shen et al. [8]	Shrinkage	ANN+GA
Altan [9]	Shrinkage	ANN
Li et al. [10]	Volumetric shrinkage Warpage	RSM NSGA-II
Song et al. [2]	Volumetric shrinkage Warpage	SVM-BP-GA
Yin et al. [11]	Warpage	BPNN
Chen et al. [12]	Warpage	Linear regression
Bensingh et al. [13]	Warpage Shrinkage Von Mises Stress	ANN+PSO
Oliaei et al. [14]	Warpage Shrinkage	ANN
Singh et al. [15]	Warpage Cycle time	Linear regression
Abdul et al. [4]	Shrinkage	ANN
Wang et al. [5]	Warpage	Kriging model
Gao and Wang [6]	Warpage	Kriging model
Ahmed et al. [16]	Warpage	Ensemble algorithm
Kumar et al. [17]	Warpage Shrinkage	PSO
Hidayah et al. [18]	Warpage	GA
Chen et al. [19]	Dimensional variation	Regression ANN
Reddy et al. [20]	Warpage	ANN SVM
Liao et al. [21]	Warpage Shrinkage	BPNN
Chiang et al. [22]	Warpage Shrinkage	RSM
Heidari et al. [23]	Warpage Shrinkage	RBF
Fernandez et al. [24]	Shrinkage Part distortion	Linear regression BPNN
Li et al. [25]	Warpage	BPNN
Yang et al. [26]	Shrinkage Warpage	BPNN
Hadler et al. [27]	Shrinkage Warpage	Linear regression

(Continued)

TABLE 4.1 (CONTINUED)

Summary of predictive modelling research

Author	Predicted Outputs	Predictive Model
Chen et al. [12]	Shrinkage Warpage	Linear regression
Hwang et al. [28]	Warpage	ANN
Kramschuster et al. [29]	Shrinkage Warpage	Linear regression
Yin et al. [30]	Warpage	Kriging model
Kang et al. [31]	Warpage	Kriging meta-model

Of the majorly used intelligent models in prediction and control of major injection molded product defects, the use of ANN stands out both in process parameter optimization and quality prediction while other models such as GA, Kriging model, RSM and other hybrid models are majorly used for optimization with minimal application to prediction and control. ANN enjoys a wide use in defects prediction models due to its precise nature and capability to linearize complex non-linear systems. However, ANN models are 'black-box' models with complex structures and hidden layers which may not give clear information about the interactions within the system and transparency in prediction. ANN therefore has limited ability to identify causal relationships and inference in that statistical relationship between input and output that is heavily employed by ANN does not imply causality.

Therefore, fuzzy logic based model is superior to neural networks in identification of possible causal relationships. With fuzzy logic model, one can clearly determine the inputs that are most strongly predictive of an output from the defined rules. In the actual injection molding environment where there are numerous input parameters each with specific contribution to the output quality indices, information about the right combinations of parameters alone is not satisfactory. It is imperative to identify the causal relationships for effective monitoring and control. This can be achieved by predictive modelling using fuzzy inference system.

Injection molded product defects predictive modelling in injection molding through fuzzy logic has not been widely explored in comparison to the other intelligent models. This is despite the many strengths that fuzzy logic possesses in terms its capability to develop accurate predictive models even with less amount of data which suits the real injection molding environment where the number of experimental trials carried out may be limited due to time and cost implications. Fuzzy logic also incorporates the designer's expertise during the rule definition and this would suit well the actual injection molding environment where various specific rules describing the effects of process parameters to the responses can be incorporated onto the model. Fuzzy logic is also capable of handling imprecision in process parameters such as temperatures and pressures which are normally defined imprecisely in terms of ranges in most molding environments such that a small change in temperature and pressure should not result to larger changes in the responses.

Moreover, most previous defects predictive modelling studies considered single stage design of experiment for determination of model training data. Records of process parameter screening to obtain the most significant parameters affecting the defects to be used for predictive modelling is unclear. Other studies that carried out parameter screening based them entirely on mains effects without considering parameter interaction effects and their contributions to the defects. The use of single stage design of experiment or process parameter screening without interaction effects may result to wrong statistical inferences and biased predictions thereby lowering the model performance. There is therefore a clear need to explore on multi-stage DOEs to enhance a thorough parameter space exploration during the screening stage and establish both the mains effect and critical interactions between process parameters and their effects to the defects.

4.3 Fuzzy Logic for Injection Molding

Fuzzy logic systems employs the use of fuzzy sets, linguistic variables, fuzzy rules, fuzzy inference system, possibility distributions and membership functions for system design and decision making. Fuzzy sets are classes of objects with varying degrees of membership ranging between 0 and 1. Other expert systems are capable of making precise decision at every stage whereas fuzzy systems are capable of retaining information concerning uncertainty and draws a precise decision in the final stage [32]. The use of linguistic rules in fuzzy logic provides a better understanding of each design characteristic alongside making the design tools more intuitive. Fuzzy logic system mainly consists of three conceptual components such as fuzzification, definition of expert rules and defuzification.

The selection of membership function types and ranges affects the accuracy of fuzzy logic predictive models. The types and ranges of membership functions can be determined through intuition, inference, rank ordering or using intelligent algorithms [33]. Upon fuzzy logic model development, tuning of the model can be carried out to optimize the membership function parameters and rules using various optimization algorithms. Tuning entails the use of intelligent algorithms to adjust membership function parameters and rules to enhance the performance of a fuzzy inference system [34]. Pattern search is such an algorithm that is a derivative-free optimization technique and explores a search space through iterative probing of various points to determine the optimum point [35]. Pattern search is a local optimization algorithm and does well for tuning models with small parameter tuning ranges in comparison to global optimization algorithms such as Genetic Algorithm which does well for large parameter tuning ranges [35].

Fuzzy logic has received a wide range of applications in the design of expert systems owing to its robust nature, application for practical purposes to prediction and control and its capability to work with smaller quantities of data. Robustness defines the ability of fuzzy logic systems to be designed in a wide variety of ways as a result of a wide range of available membership functions to choose from, wide range of fuzzification methods, defuzification methods and wide range of methods of rule definitions and rule weightings. Fuzzy logic expert systems do not need larger sets of data to obtain accurate predictive models. The accuracy of the predictive models depends on the expertise of the designer in terms of rule definition and choice of membership functions [33].

Fuzzy logic systems have been used in plastic injection molding for various purposes majorly on process parameter optimization and quality assessment. Moayyedian et al. [36] utilized fuzzy logic for quality evaluation and optimization of process parameters to achieve a higher moldability index that was given as a function of short shot, shrinkage rate and warpage. This study however did not develop predictive models for the moldability index or the defects. Chaves et al. [37] utilized fuzzy logic for quality assessment of injection molded parts to prevent flash and burn mark defects and remarkable optimal results. Other studies such as Chiang and Chang [38] utilized fuzzy logic for optimization of process parameters for warpage reduction while Hu and Wu [39] utilized fuzzy logic in the design of a fuzzy expert controller for barrel temperature control in injection molding machine.

4.4 Defects Prediction Based on Fuzzy Logic and Pattern Search

4.4.1 Study Design

A case study was carried out to model and predict various plastic injection molding defects using fuzzy logic and pattern search optimization algorithm. A weighted defect index was computed as outlined in Section 3.5.5. The study used a combination of design of experiment, CAE simulations and intelligent algorithms for the weighted defect index predictive modelling. The numerical data used for predictive model training and validation were obtained using Moldex3D® R22 CAE software and fuzzy predictive model developed and tuned using MATLAB® Fuzzy Logic Designer Toolbox.

A two-stage design of experiment was carried out to enhance process parameter screening. This included a factorial design of experiment and Taguchi design of experiment. To establish the mains effects, interaction effects and carry out process parameter screening based on the established and computed effect sizes, a fractional factorial design of experiment was carried out as outlined in Section 3.5.1. The results from this design of experiment were used for identification of process parameter mains effects, identification and computation of interaction effects sizes and selection of the most significant parameters for model development. Taguchi design of experiment was used for generation of model training and validation data based on CAE modelling.

Fuzzy expert models were developed to predict the values of warpage and shrinkage defects at given process parameter inputs. A forward mapped Mamdani Fuzzy Inference System (FIS) was developed using MATLAB® Fuzzy Logic Designer and utilizing Gaussian membership function type. The main components of the fuzzy logic predictive model developed were the fuzzifier, linguistic rule base, inference engine and defuzzifier.

4.4.2 Process Parameter Screening

Selection of the most significant process parameters to be used in the second design of experiment stage was done based on ANOVA results from the factorial design of experiment stage outlined in Section 3.5.5. Table 4.2 outlines a section of the ANOVA results from the factorial design of experiment stage.

TABLE 4.2

ANOVA for factorial DOE stage

Source	DF	Seq SS	Contribution (%)	Adj SS	Adj MS	F-Value	P-Value
Melt Temp	1	0.05585	3.27	0.055852	0.055852	50.03	0.000
Mold Temp	1	0.00033	0.02	0.000330	0.000330	0.30	0.592
Inj. Press.	1	0.58190	34.06	0.581897	0.581897	521.19	0.000
Pack Time	1	0.10976	6.42	0.109765	0.109765	98.31	0.000
Pack Press.	1	0.89638	52.47	0.896377	0.896377	802.86	0.000
Cool Time	1	0.01631	0.95	0.016315	0.016315	14.61	0.001
Melt Temp*Pack Press.	1	0.01612	0.94	0.016116	0.016116	14.43	0.001
Pack Time*Pack Press.	1	0.00607	0.36	0.006075	0.006075	5.44	0.029
Error	23	0.02568	1.50	0.025679	0.001116		
Total	31	1.70841	100.00				

The results indicate mold temperature having no significant mains and interaction effect on the weighted defect index.

Therefore, for the second design of experiment and subsequent predictive model development, five factors including melt temperature, injection pressure, packing time, packing pressure and cooling time were used.

4.4.3 Taguchi Design of Experiment

Upon process parameter screening, five of the six factors including melt temperature, injection pressure, packing time, packing pressure and cooling time were selected and utilized in the second stage design of experiment. Three levels of factor application have been commonly used in previous defect predictive studies [40, 41]. However, to enhance a more robust design and increase the predictive ability of the fuzzy inference model, five levels were used for each input parameter input. Taguchi design was used for five input variables each at five levels and yielded an L25 orthogonal array. Numerical warp simulations were carried out at a constant mold temperature of 35°C which is the average recommended temperature for use with the material and the numerical data validated through comparison of mains effect trends against those reported by [42]. Values of four major defects that is shrinkage, warpage, short shot and sink marks were obtained at each run and the weighted defect index computed based on Equations (3.3) and (3.4) from Section 3.5.5 of the third chapter. With the general objective of the modelling being to minimize the defect index, a smaller the better signal to noise ratio was calculated based on Equation (4.1) [9].

$$SN = -10 \log \left[\frac{1}{n} \sum_{i=1}^{n} y_i^2 \right] \tag{4.1}$$

In addition to the predictive model development data obtained from Taguchi design, an additional simulation was carried out at different input levels that would be used for testing the performance and validation of the developed predictive models. Seven sets of input parameter combinations were randomly generated and used to perform numerical simulations. The distribution of the data was widened in order to accurately represent the problem space and prevent further problems that would arise due to validation data set bias. Table 4.3 shows the results of the numerical simulations.

Table 4.4 shows the results obtained from the numerical simulations of the validation data set. Figure 4.1 is a layout of the mains effect trends for shrinkage obtained from the simulations and experiment for A - Melt Temperature, B - Injection Pressure, C - Packing Pressure, D -Pack/Hold Time and E - Cooling Time. It shows the general comparative effects of the changes in process parameters to shrinkage defect.

TABLE 4.3

Weighted Defect Index Results Based on Taguchi L25 Design

Run	Melt Temp.	Injection Pressure	Pack Pressure	Pack Time	Cool Time	Defect Index
1	200	100	100	2	5	0.976
2	200	150	150	3	8	0.678
3	200	200	200	4	10	0.397
4	200	250	250	5	12	0.255
5	200	300	300	6	15	0.070
6	210	100	150	4	12	0.741
7	210	150	200	5	15	0.448
8	210	200	250	6	5	0.322
9	210	250	300	2	8	0.212
10	210	300	100	3	10	0.517
11	220	100	200	6	8	0.584
12	220	150	250	2	10	0.449
13	220	200	300	3	12	0.213
14	220	250	100	4	15	0.491
15	220	300	150	5	5	0.425
16	230	100	250	3	15	0.523
17	230	150	300	4	5	0.303
18	230	200	100	5	8	0.581
19	230	250	150	6	10	0.373
20	230	300	200	2	12	0.422
21	240	100	300	5	10	0.402
22	240	150	100	6	12	0.610
23	240	200	150	2	15	0.592
24	240	250	200	3	5	0.401
25	240	300	250	4	8	0.193

TABLE 4.4

Model Validation Dataset

Run	Melt Temp.	Injection Pressure	Pack Pressure	Pack Time	Cool Time	Defect Index
1	207	120	108	2	6	0.904
2	222	285	108	6	11	0.425
3	203	134	174	3	6	0.685
4	207	266	240	5	13	0.281
5	207	226	223	4	11	0.359
6	223	158	240	2	11	0.381
7	236	186	174	2	15	0.596

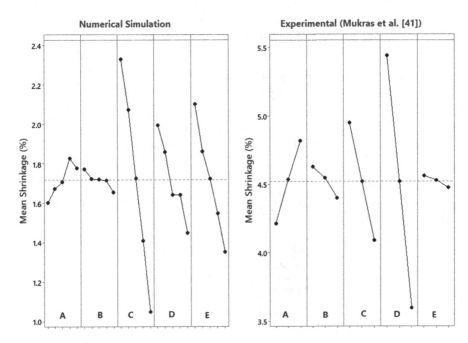

FIGURE 4.1
Comparative mains effect plots from second stage numerical simulation and experiment.

As obtained from the fractional factorial design, there was a similarity in trends for both the warpage and shrinkage plots. A general increase in melt temperature and injection pressure increased shrinkage whereas an increase in packing pressure, packing time and cooling time decreased shrinkage. Similarity in the data trends was satisfactory and hence the data was used for predictive model development.

4.4.4 Fuzzification

Through fuzzification, crisp input variables of five parameters namely melt temperature, injection pressure, packing pressure, packing time and cooling time were mapped onto fuzzy variables by expression as membership functions. Five membership functions were used for each of the five input variables. Linguistic labels chosen for each of the input membership functions were; Very Low (VL), Low (L), Medium (M), High (H) and Very High (VH). After iteratively trying different membership function types and their associated effects to the defect responses as illustrated on Figure 4.2, Gaussian membership function type was deemed suitable and thus used in fuzzification of the input variables.

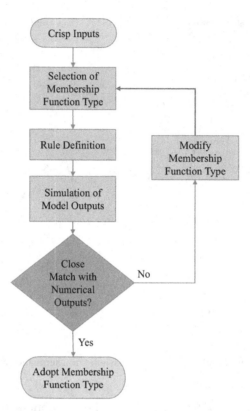

FIGURE 4.2
Membership function type selection method.

The degrees of membership in Gaussian membership functions were expressed as given in Equations (4.2) [43].

$$f(x; \sigma, \mu) = e^{\frac{-(x-\mu)^2}{2\sigma^2}} \tag{4.2}$$

Where σ is the standard deviation of the membership class representing its width property, μ is the mean value of the membership class corresponding to the peak value while x is any arbitrary point whose membership value is to be determined.

Parameters of the membership functions were defined based on their levels from the design of experiment. Figures 4.3 and 4.4 illustrates the Gaussian membership functions plots for two of the inputs namely injection pressure and cooling time.

An injection pressure of 125 MPa has a 0.5 degree of membership on the Very Low membership function and 0.5 membership of the Low membership

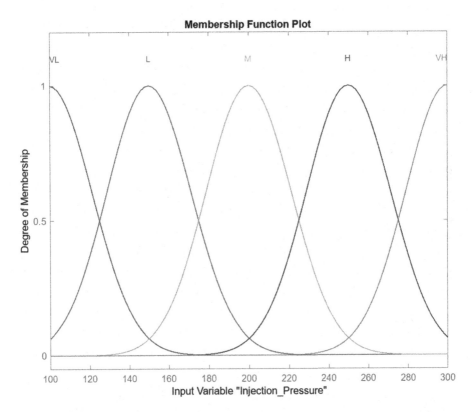

FIGURE 4.3
Membership plot for injection pressure input values.

function. For cooling time, a membership function 'Medium' encompasses cooling time ranges of 7.5s to 12.5s with the maximum membership value of 1 at 10s. For all the input variables, the peak values of the five membership functions correspond to the five levels of application of the variables adopted during the Taguchi design of experiment.

4.4.5　Rule Base Generation

Fuzzy rule base was employed to express the relationships between input and output variables. For 5 input parameters each having 5 membership functions, a total of 3125 rules would be possible if a full factorial experiment design was used. However, since the experiment was designed using Taguchi L25 method which yielded 25 runs of simulation data, these runs of data were used for rule generation. One of the strengths of fuzzy logic predictive model is the lack of data intensity in comparison with the other predictive models. Various studies have successfully developed accurate

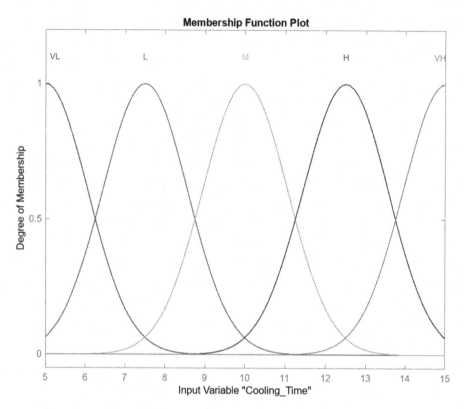

FIGURE 4.4
Membership plot of cooling time input values.

fuzzy predictive models based solely on data obtained from Taguchi design of experiments [44–47].

Therefore, a set of 25 *IF-THEN* rules (Table 4.5) governing the mapping of inputs into outputs were formulated through expert knowledge based on the 25 runs of data obtained from Taguchi based numerical simulations. The study assigned equal weights of 1 to each rule. This study did not take into consideration the possible use of multiple sub-fuzzy models as they would isolate the effects of individual process parameters, something which is not representative of plastic injection molding process that involves simultaneous interaction of all the process parameter to influence the defects. Two of the rules numbered 1 and 5 can be read as follows respectively:

IF (Melt Temperature is VL) and (Injection Pressure is VL) and (Packing Pressure is VL) and (Packing Time is VL) and (Cooling Time is VL) THEN (Weighted Defect Index is D10)

TABLE 4.5

Linguistic Variables for the FIS Rule Base

Rule No.	Melt Temp	Inj. Press.	Pack Press.	Pack Time	Cool Time	Weighted Defects Index
1	VL	VL	VL	VL	VL	D10
2	VL	L	L	L	L	D7
3	VL	M	M	M	M	D4
4	VL	H	H	H	H	D3
5	VL	VH	VH	VH	VH	D1
6	L	VL	L	M	H	D7
7	L	L	M	H	VH	D4
8	L	M	H	VH	VL	D3
9	L	H	VH	VL	L	D2
10	L	VH	VL	L	M	D5
11	M	VL	M	VH	L	D6
12	M	L	H	VL	M	D4
13	M	M	VH	L	H	D2
14	M	H	VL	M	VH	D5
15	M	VH	L	H	VL	D4
16	H	VL	H	L	VH	D5
17	H	L	VH	M	VL	D3
18	H	M	VL	H	L	D6
19	H	H	L	VH	M	D4
20	H	VH	M	VL	H	D4
21	VH	VL	VH	H	M	D4
22	VH	L	VL	VH	H	D6
23	VH	M	L	VL	VH	D6
24	VH	H	M	L	VL	D4
25	VH	VH	H	M	L	D2

IF (Melt Temperature is VL) and (Injection Pressure is VH) and (Packing Pressure is VH) and (Packing Time is VH) and (Cooling Time is VH) THEN (Weighted Defect Index is D1)

4.4.6 Defuzification

To convert fuzzy variables back to crisp values, a centroid defuzification method was used. This method calculated the center of area under the membership function and weighed the effect of each input variable towards the calculation. The study utilized ten membership functions for weighted defect index responses.

To select the number of membership functions for the output variables, the ranges of the variables obtained from simulation data were considered.

The lowest value of weighted defect index obtained was 0.07 with the highest value being 0.976 giving a universe of discourse of 0.07 to 0.976. The number of fuzzy sets and hence the membership functions selected should not be too few nor too many. The use of too many membership functions may have merits such as more detailed output variable representation and increased model accuracy which may also come at a cost of data overfitting and model complexity.

It is therefore necessary to strike a balance in selection of the number of membership functions to enhance accuracy while also lowering the chances of overfitting. If five membership functions were to be used considering the universe of discourse as the output variable ranges, the intervals of each function class would be 0.2 and up to seven data points would lie in a single membership class. This would significantly lower the predictive capability of the model as a result of the wider membership class interval. Having ten membership functions would yield a membership class interval of 0.1 and only a maximum of three data points would lie in a single membership class. The study therefore utilized ten membership functions for the output response. The membership functions were specified by parameters such as the standard deviation and the mean.

Figure 4.5 shows Gaussian membership function plots for weighted defect index response with 10 membership function classes. All the membership function classes were symmetrically placed, had a standard deviation of 0.0425 while the mean of each class is defined as the peak value of the given class.

4.4.7 Fuzzy Inference System Structure

Figure 4.6 illustrates the developed FIS model structure with five inputs and one output and designed with Gaussian membership functions for input and output variables.

Parameter causality was easily identified from the rules interface of the developed models as illustrated on Figure 4.7. It can be deduced from the interface that lower values of melt temperature are associated with lower values of shrinkage and higher values of warpage whereas lower values of packing pressure are associated with higher shrinkages and warpages.

Linguistic rule base representation from the FIS rule inference directly expresses the relationships among variables and hence facilitates an easy interpretation of how the model makes decisions.

In contrast to neural networks with complex structure with multiple hidden layers, the linguistic rule base representation from the FIS rule inference directly expresses the relationships among variables and hence facilitates an easy understanding and interpretation of how the model makes decisions. This could form a basis for on-line injection molding process optimization.

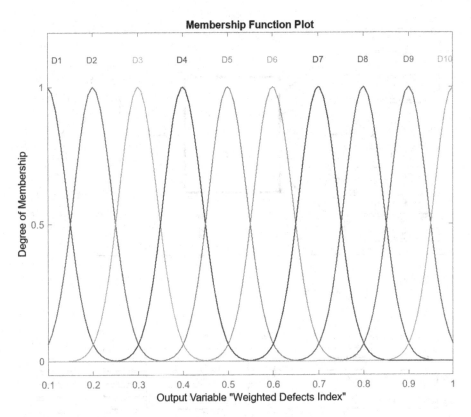

FIGURE 4.5
Membership functions for the weighted defect index.

For instance, whenever a change in the ambient room temperature affects the melt temperature of the material and hence the cooling time, with the aid of the model rule inference, the other parameters such as injection pressure, packing pressure and packing time can be adjusted to reduce the resulting defect index or maintain them at the required rates.

4.4.8 Fuzzy Logic Model Plots

Figure 4.8 illustrate a 3D surface plots from the developed FIS showing the relationship between packing pressure and injection pressure to the defects index output. The lowest defect indices are obtained at higher values of packing pressure and injection pressure whereas the highest values of defect index are obtained at lower values of injection pressure and packing pressure. This relationship was similar to that obtained statistically illustrated in Figure 4.9.

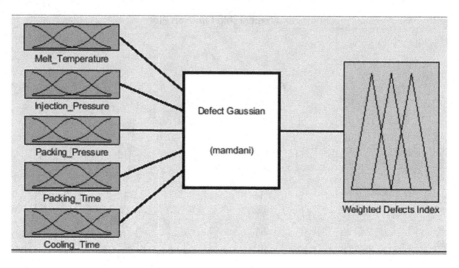

FIGURE 4.6
Fuzzy predictive model structure.

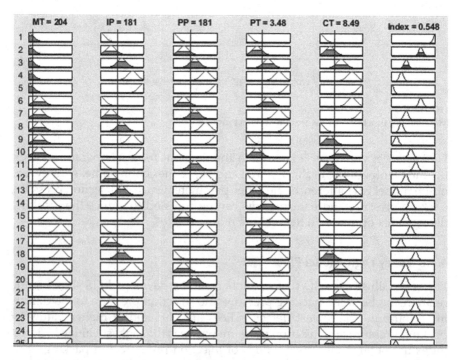

FIGURE 4.7
Fuzzy predictive model rule base graphical interface.

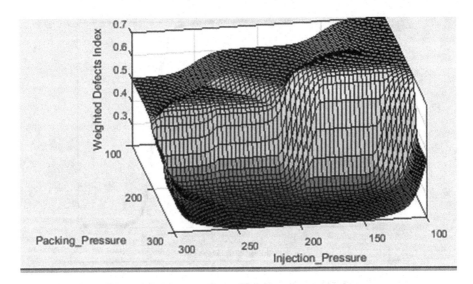

FIGURE 4.8
FIS relationship between pack pressure and injection pressure.

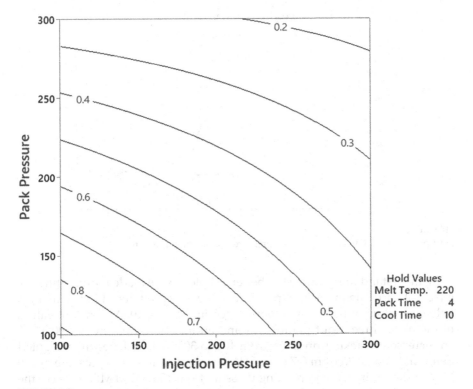

FIGURE 4.9
Statistical relationship between pack pressure and injection pressure.

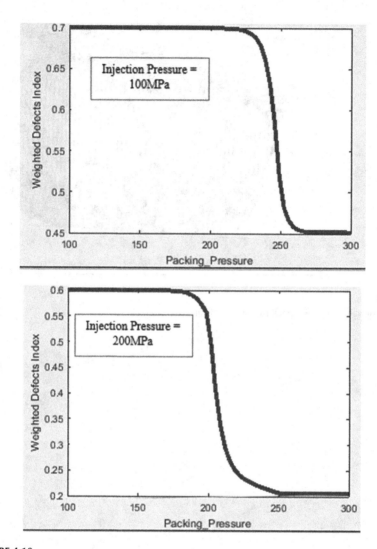

FIGURE 4.10
2D representation of the effects of pack pressure and injection pressure.

To understand the relationship better, 2D plots were made representing the effects of variation in packing pressure to the weighted defect index at given values of injection pressure as represented in Figure 4.10. At constant values of the other three input parameters and an injection pressure of 100 MPa, an increase in packing pressure from 100 to 300 MPa reduces the weighted defect index by 0.25 from 0.7 to 0.45. Whereas, at an injection pressure of 200 MPa, a similar increase in packing pressure from 100 to 300 MPa reduces the weighted defect index by 0.4 from 0.6 to 0.2. This implies that the developed

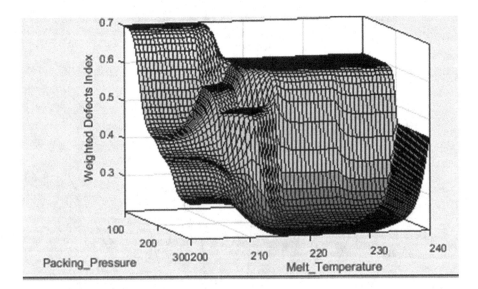

FIGURE 4.11
FIS relationship between pack pressure and melt temperature.

fuzzy inference model captured the complex relationship in terms of interactions between the two process parameters and the effects of such interactions on the output. That is, the mains effect of packing pressure on defect index is magnified at higher values of injection pressure and diminished at lower values of injection pressure and vice versa.

Figure 4.11 illustrates the 3D surface plot indicating the relationship between packing pressure, melt temperature and weighted defect index. Packing pressure accounts for the biggest change in the defect index with the highest defect index obtained at the lowest values of packing pressure and melt temperature while the lowest defect index is obtained at the highest value of packing pressure and the lowest value of melt temperature. This relationship was similar to that obtained statistically and illustrated in Figure 4.12.

The representation of the relationship between the melt temperature and packing pressure in two dimensions as given on Figure 4.13 also indicated an interaction effect between the process parameters. That is, at a melt temperature setting of 200°C and constant values of the other three input parameters, an increase in packing pressure from 100 MPa to 300 MPa decreases the weighted defect index by 0.4 from 0.7 to 0.3. However, at a melt temperature setting of 240°C, a similar increase in packing pressure decreases the defect index by 0.15 from 0.6 to 0.45 with an altered nature of decrease as represented by the curve. This relationship indicates that fuzzy logic models captures the interaction effects among process parameters.

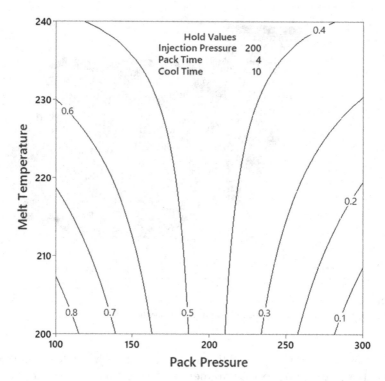

FIGURE 4.12
Statistical relationship between pack pressure and melt temperature.

4.4.9 Model Tuning and Validation

This study determined the parameters of the fuzzy model membership functions through intuition and further carried out tuning of the parameters using a pattern search optimization technique [35]. Tuning of fuzzy model parameters was carried out based on the data set of the simulation results used for rule formulation. The tuning was based on a default cost function obtained as the root mean square error between the actual outputs and the initial fuzzy predicted outputs.

The convergence criteria used for the tuning was the distance metric (RMSE) of data sets in Table 4.3 used for development of the fuzzy inference system. The objective function $f(x)$ to be minimized was calculated for the FIS based on the differences between the reference outputs obtained from simulations and FIS predicted outputs. For twenty five data sets, $f(x)$ was calculated using Equation (4.3).

$$f(x) = \sqrt{\frac{1}{25} \sum_{i=1}^{25} \left(\left(y_{i,\text{ref}} - y_{i,\text{pred}} \right)^2 \right)} \qquad (4.3)$$

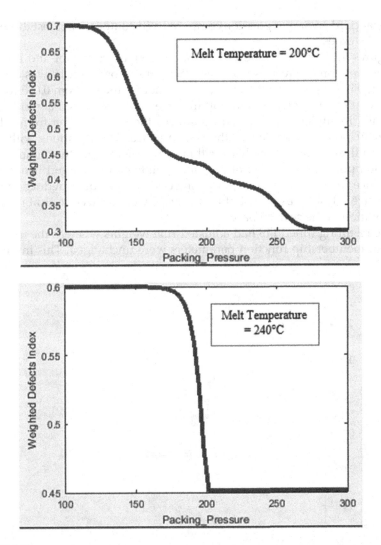

FIGURE 4.13
2D representation of the effect of pack pressure and melt temperature.

Where $y_{i,\text{ref}}$ is the reference defect index output for the ith data point and $y_{i,\text{pred}}$ is the predicted defect index output for the ith data point. With the objective function in place, the optimization algorithm iteratively adjusted the various fuzzy inference system parameters such as input membership function parameters, output membership function parameters and rules until the objective function converged to a minimum solution. Upon tuning, the initial models and tuned models were tested against a validation data set and performance metrics such as RMSE, coefficient of determination and standard error of regression computed and compared against those from related studies.

Figures 4.14 and 4.15 shows the tuning convergence plot and a sample of the convergence results from the FIS tuning. Convergence was achieved after 160 iterations with the objective function reducing from 0.035 to 0.018. The iterations with successful poll indicates that the search algorithm successfully identified a point that minimizes the cost function from a list of potential points generated in the search space. The iterations with refine mesh indicate that the search algorithm could not identify the minimum cost function point from the list of potential points generated and hence it had to reduce the search space exploration steps sizes through reduction of the mesh size. Mesh sizes from 4 of up to 0.0078 were utilized indicating a finer exploration of the search space.

The resulting tuned FIS had adjusted rule weights whereas the input and output membership function parameters were unchanged. This implied an

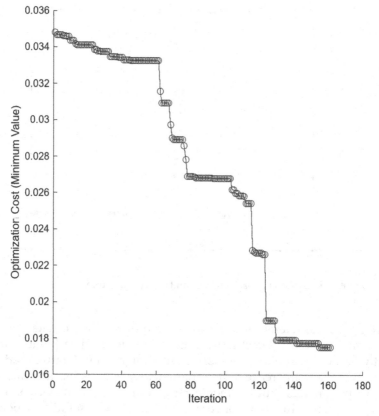

FIGURE 4.14
Tuning convergence plot.

Convergence Results ×

System: Defect Index

Iteration	Func-count	f(x)	MeshSize	Method
0	1	0.03499	1	
1	46	0.03482	2	Successful Poll
2	67	0.03468	4	Successful Poll
3	67	0.03468	2	Refine Mesh
4	98	0.03468	4	Successful Poll
5	98	0.03468	2	Refine Mesh
6	129	0.0346	4	Successful Poll
7	129	0.0346	2	Refine Mesh
8	160	0.03458	4	Successful Poll
9	160	0.03458	2	Refine Mesh
10	226	0.03436	4	Successful Poll
11	226	0.03436	2	Refine Mesh
12	307	0.03436	1	Refine Mesh
13	372	0.03416	2	Successful Poll
14	399	0.03412	4	Successful Poll
15	399	0.03412	2	Refine Mesh
16	478	0.03412	4	Successful Poll
17	478	0.03412	2	Refine Mesh
18	556	0.03412	4	Successful Poll
19	556	0.03412	2	Refine Mesh
20	638	0.03412	1	Refine Mesh
141	13303	0.01791	0.5	Refine Mesh
142	13365	0.01774	1	Successful Poll
143	13540	0.01774	2	Successful Poll
144	13568	0.01774	4	Successful Poll
145	13568	0.01774	2	Refine Mesh
146	13656	0.01774	1	Refine Mesh
147	13865	0.01774	0.5	Refine Mesh
148	13966	0.01774	1	Successful Poll
149	14176	0.01774	0.5	Refine Mesh
150	14265	0.01774	1	Successful Poll
151	14476	0.01774	0.5	Refine Mesh
152	14719	0.01774	1	Successful Poll
153	14930	0.01774	0.5	Refine Mesh
154	15185	0.01774	0.25	Refine Mesh
155	15210	0.01753	0.5	Successful Poll
156	15465	0.01753	0.25	Refine Mesh
157	15723	0.01753	0.125	Refine Mesh
158	15981	0.01753	0.0625	Refine Mesh
159	16239	0.01753	0.03125	Refine Mesh
160	16497	0.01753	0.01562	Refine Mesh
161	16755	0.01753	0.007812	Refine Mesh

Maximum number of iterations exceeded: increase options.MaxIterations.

FIGURE 4.15
Sample of the convergence results.

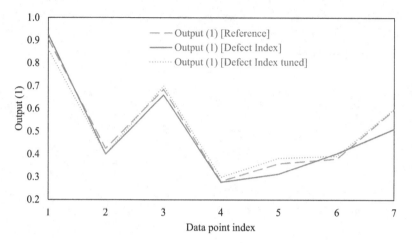

FIGURE 4.16
Comparative performance of the developed defect index FISs.

optimum definition of the input and output membership function parameters. Upon successful tuning, the performances of the initially designed
defect index FIS and tuned defect index FIS were tested against a validation
data set given in Table 4.5. Using the outputs from the validation data set as
the reference outputs, the RMSE for both the initial FIS and tuned FIS were
obtained. Figure 4.16 illustrates a comparative evaluation of the performance
of the initially developed FIS and tuned FIS against the reference numerical
outputs. In general, the tuned FIS curve was closer to the reference curve for
most data points thereby indicating an accurate prediction capability.

Figure 4.17 illustrates a comparison of the prediction errors from the initially designed FIS and the tuned FIS. Positive prediction errors indicated an
over prediction of the defect index whereas negative prediction errors indicated an under prediction. The initially designed FIS over predicted most
of the data points whereas the tuned FIS under predicted most of the data
points. The prediction errors were smaller and hence sufficient. The resulting tuned FIS had a lower value of RMSE of 0.023 compared to the initially
developed FIS that had an RMSE value of 0.04. Also, the initially developed
FIS had the largest deviations from the reference values for most of the data
points compared to the tuned FIS. Through tuning, the search algorithm captured underlying relationships in the data and modified given rule weightings to better reflect the data patterns thereby reducing model prediction
errors [34].

In addition to the RMSE, a coefficient of determination and standard error
of regression were determined for the two FISs to assess the abilities of the
models to predict the outcomes in a linear regression setting. Coefficient of

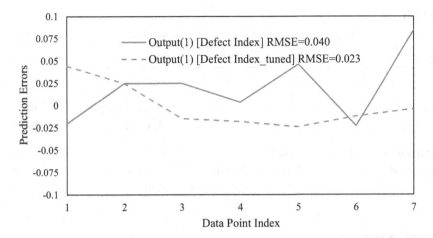

FIGURE 4.17
Comparison of the FIS prediction errors.

determination and standard error of regression were calculated at 95% confidence level. As a result of the possible non-linearity of plastic injection molding process, this study used standard error of regression metric to evaluate the performance of the predictive model alongside coefficient of determination metric.

Table 4.6 shows the performance metrics of the predictive models computed in a regression setting. The tuned FIS had lower RMSE for prediction of the defect index output indicating a good predictive capability. It also had a lower standard error of regression and higher coefficient of determination implying that the model fit the data set more closely. The general good performance of the FIS models on the validation data set was an indication of minimal chances of model overfitting to the training data set.

These performance metrics compared well with those of other predictive models based on other strategies reported in literature as summarized on Table 4.7.

TABLE 4.6

Performance Metrics of the Developed FISs

FIS	Performance Metric		
	Root Mean Square Error (RMSE)	**Standard Error of Regression (S) (%)**	**Coefficient of Determination (R²) (%)**
Defect Index	0.040	0.0412	97.25
Defect Index tuned	0.023	0.0211	99.11

TABLE 4.7

Model Performance Metrics Reported from Similar Studies

Study	Model	Response	CoD (R^2)
Paturi et al. [3]	ANN	Weld-line width	99.95%
		Sink mark depth	
Chen et al. [48]	ANN	Dimensional variation	91%
Kumar et al. [17]	PSO + Regression	Shrinkage	94.1%
		Warpage	92.4%
Hidayah et al. [18]	GA + RSM	Warpage	98.1%
Ahmed et al. [16]	Random forest algorithm	Warpage	96.75%
	Gradient boosted regression	Warpage	90.63%

4.5 Conclusion

A multi-stage design of experiment in predictive modelling of various injection molding defects helps to improve the overall model performance. Consideration of process parameter interactions as well as parameter mains effects in the parameter screening stage helps to achieve a more accurate and robust predictive model. Interactions accounts for the combined effects of variables and thereby allowing the model to capture how the individual effects of a certain parameter depends on the levels of the others. This also helps in capturing non-linear relationships between the input and output variables which is a major feature of injection molding process.

The integration of fuzzy logic rule base systems with intelligent optimization algorithms for predictive modelling of shrinkage and warpage defects in plastic injection molding can provide a yardstick to the defects minimization and quality control. As a result of the variability of the injection molding process with regard to the conditions of the molding environment, a fuzzy logic model can be designed incorporating expert experience on the relationships among input and output variables. In addition to the available historical data, an injection molding expert can incorporate his experience and expertise in development of a rule base which is specific to the given molding environment.

In the data collection through Taguchi design of experiment for model development, the use of more levels of the input facilitates the use of more input membership functions which result to a robust model capable of capturing finer variations in the input parameters. The expression of the relationships between input parameters and output responses and their representation in fuzzy logic rule inference provides an avenue for establishing the causal relationships among parameters. In weighted defect index modelling through fuzzy logic, knowledge of the causal relationships from the fuzzy rule inference provides more insight into the system and serves as a decision tool for effective process monitoring and control.

Tuning of expert designed fuzzy inference systems using intelligent algorithms helps to improve the predictive ability of the models and hence would lead to enhanced performance. Through tuning, an intelligent algorithm identifies complex patterns within the FIS and adjusts the FIS parameters to better match the patterns. Tuned fuzzy logic models performs better regardless of the amount of data available for design.

Therefore, the integration of multi-stage design of experiments, CAE and intelligent algorithms in predictive modelling of plastic injection molded product defects represents a significant stride towards quality control. Predictive modelling of the defects empowers manufacturers to identify and control the defects prior to occurrence. Design of experiments provides a systematic approach to exploring the process through variation of a number of factors and determining the impact of the variation to the final product. This provides historical information which becomes the basis by which the intelligent predictive models are based.

References

[1] N. Y. Zhao, J. Y. Lian, P. F. Wang, and Z. Bin Xu, "Recent progress in minimizing the warpage and shrinkage deformations by the optimization of process parameters in plastic injection molding: A review," *International Journal of Advanced Manufacturing Technology*, vol. 120, no. 1–2, pp. 85–101, Feb. 10, 2022. doi: 10.1007/s00170-022-08859-0

[2] Z. Song, S. Liu, X. Wang, and Z. Hu, "Optimization and prediction of volume shrinkage and warpage of injection-molded thin-walled parts based on neural network," *Int J Adv Manuf Tech*, vol. 109, no. 3–4, pp. 755–769, Jul. 2020. doi: 10.1007/s00170-020-05558-6

[3] U. M. R. Paturi, S. K. P. Kolluru, and S. D. S. A. Kalvakolanu, "Prediction of weld-line width and sink-mark depth of plastic injection moulded parts using neural networks," *Mater. Today Proc.*, Mar. 2023. doi: 10.1016/j.matpr.2023.02.295

[4] R. Abdul, G. Guo, J. C. Chen, J. Jung, and W. Yoo, "Shrinkage prediction of injection molded high density polyethylene parts with taguchi/artificial neural network hybrid experimental design," *Int. J. Interact. Des. Manuf.*, no. 0123456789, 2019. doi: 10.1007/s12008-019-00593-4

[5] X. Wang, J. Gu, C. Shen, and X. Wang, "Warpage optimization with dynamic injection molding technology and sequential optimization method," *Int. J. Adv. Manuf. Technol.*, vol. 78, no. 1–4, pp. 177–187, Apr. 2015. doi: 10.1007/s00170-014-6621-x

[6] Y. Gao and X. Wang, "An effective warpage optimization method in injection molding based on the Kriging model," *Int. J. Adv. Manuf. Technol.*, vol. 37, no. 9, pp. 953–960, 2008. doi: 10.1007/s00170-007-1044-6

[7] E. Hakimian and A. B. Sulong, "Analysis of warpage and shrinkage properties of injection-molded micro gears polymer composites using numerical simulations assisted by the Taguchi method," *Mater. Des.*, vol. 42, pp. 62–71, Dec. 2012. doi: 10.1016/J.MATDES.2012.04.058

[8] C. Shen, L. Wang, and Q. Li, "Optimization of injection molding process parameters using combination of artificial neural network and genetic algorithm method," *J. Mater. Process. Technol.*, vol. 183, no. 2–3, pp. 412–418, Mar. 2007. doi: 10.1016/j.jmatprotec.2006.10.036

[9] M. Altan, "Reducing shrinkage in injection moldings via the Taguchi, ANOVA and neural network methods," *Mater. Des.*, vol. 31, no. 1, pp. 599–604, Jan. 2010. doi: 10.1016/J.MATDES.2009.06.049

[10] K. Li, S. Yan, Y. Zhong, W. Pan, and G. Zhao, "Multi-objective optimization of the fiber-reinforced composite injection molding process using Taguchi method, RSM, and NSGA-II," *Simul. Model. Pract. Theory*, vol. 91, pp. 69–82, Feb. 2019. doi: 10.1016/J.SIMPAT.2018.09.003

[11] F. Yin, H. Mao, L. Hua, W. Guo, and M. Shu, "Back Propagation neural network modeling for warpage prediction and optimization of plastic products during injection molding," *Mater. Des.*, vol. 32, no. 4, pp. 1844–1850, Apr. 2011. doi: 10.1016/J.MATDES.2010.12.022

[12] S. C. Chen, B. L. Tsai, C. C. Hsieh, N. T. Cheng, E. N. Shen, and C. Te Feng, "Prediction of part shrinkage for injection molded crystalline polymer via cavity pressure and melt temperature monitoring," *Appl. Sci.*, vol. 13, no. 17, p. 9884, Aug. 2023. doi: 10.3390/app13179884

[13] R. J. Bensingh, R. Machavaram, S. R. Boopathy, and C. Jebaraj, "Injection molding process optimization of a bi-aspheric lens using hybrid artificial neural networks (ANNs) and particle swarm optimization (PSO)," *Meas. J. Int. Meas. Confed.*, vol. 134, pp. 359–374, Feb. 2019. doi: 10.1016/j.measurement.2018.10.066

[14] E. Oliaei et al., "Warpage and shrinkage optimization of injection-molded plastic spoon parts for biodegradable polymers using Taguchi, ANOVA and artificial neural network methods," *J. Mater. Sci. Technol.*, vol. 32, no. 8, pp. 710–720, Aug. 2016. doi: 10.1016/j.jmst.2016.05.010

[15] G. Singh, M. K. Pradhan, and A. Verma, "Multi response optimization of injection moulding process parameters to reduce cycle time and warpage," *Mater. Today: Proc.*, Jan. 2018, pp. 8398–8405. doi: 10.1016/j.matpr.2017.11.534

[16] T. Ahmed, P. Sharma, C. L. Karmaker, and S. Nasir, "Warpage prediction of Injection-molded PVC part using ensemble machine learning algorithm," *Mater. Today Proc.*, vol. 50, pp. 565–569, Jan. 2022. doi: 10.1016/J.MATPR.2020.11.463

[17] S. Kumar, A. K. Singh, and V. K. Pathak, "Modelling and optimization of injection molding process for PBT/PET parts using modified particle swarm algorithm," *Indian J. Eng. Mater. Sci.*, vol. 27, no. 3, pp. 603–615, 2020. doi: 10.56042/ijems.v27i3.45057

[18] M. H. N. Hidayah, Z. Shayfull, N. Z. Noriman, M. Fathullah, R. Norshahira, and A. T. N. A. Miza, "Optimization of warpage on plastic part by using genetic algorithm (GA)," *AIP Conf. Proc.*, vol. 2030, no. November, 2018. doi: 10.1063/1.5066804

[19] J. C. Chen, G. Guo, and Y. H. Chang, "Intelligent dimensional prediction systems with real-time monitoring sensors for injection molding via statistical

regression and artificial neural networks," *Int. J. Interact. Des. Manuf.*, vol. 17, no. 3, pp. 1265–1276, Jun. 2023. doi: 10.1007/s12008-022-01115-5

[20] B. S. Reddy, J. S. Kumar, V. K. Reddy, and G. Padmanabhan, "Application of soft computing for the prediction of warpage of plastic injection molded parts," *J. Eng. Sci. Technol. Rev.*, vol. 2, no. 1, pp. 56–62, 2009. doi: 10.25103/jestr.021.11

[21] S. J. Liao, W. H. Hsieh, J. T. Wang, and Y. C. Su, "Shrinkage and warpage prediction of injection-molded thin-wall parts using artificial neural networks," *Polym. Eng. Sci.*, vol. 44, no. 11, pp. 2029–2040, 2004. doi: 10.1002/pen.20206

[22] K. T. Chiang and F. P. Chang, "Analysis of shrinkage and warpage in an injection-molded part with a thin shell feature using the response surface methodology," *Int. J. Adv. Manuf. Technol.*, vol. 35, no. 5–6, pp. 468–479, 2007. doi: 10.1007/s00170-006-0739-4

[23] B. Shiroud Heidari, A. H. Moghaddam, S. M. Davachi, S. Khamani, and A. Alihosseini, "Optimization of process parameters in plastic injection molding for minimizing the volumetric shrinkage and warpage using radial basis function (RBF) coupled with the k-fold cross validation technique," *J. Polym. Eng.*, vol. 39, no. 5, pp. 481–492, May 2019. doi: 10.1515/polyeng-2018-0359

[24] A. Fernández, I. Clavería, C. Pina, and D. Elduque, "Predictive methodology for quality assessment in injection molding comparing linear regression and neural networks," *Polymers (Basel)*, vol. 15, no. 19, p. 3915, Oct. 2023. doi: 10.3390/polym15193915

[25] Q. Li, L. Li, X. Si, and W. Rongji, "Modeling the effect of injection molding process parameters on warpage using neural network theory," *J. Macromol. Sci. Part B Phys.*, vol. 54, no. 9, pp. 1066–1080, Sep. 2015. doi: 10.1080/00222348.2015.1068680

[26] K. Yang, L. Tang, and P. Wu, "Research on optimization of injection molding process parameters of automobile plastic front-end frame," *Adv. Mater. Sci. Eng.*, vol. 2022, pp. 1–18, 2022. doi: 10.1155/2022/5955725

[27] N. Hadler Marins, F. Bier de Mello, R. Marques e Silva, and F. Aulo Ogliari, "Statistical approach to analyze the warpage, shrinkage and mechanical strength of injection molded parts," *Int. Polym. Process.*, vol. 31, no. 3, pp. 376–384, Jul. 2016. doi: 10.3139/217.3219

[28] S. Hwang and J. Kim, "Injection mold design of reverse engineering using injection molding analysis and machine learning," *J. Mech. Sci. Technol.*, vol. 33, no. 8, pp. 3803–3812, Aug. 2019. doi: 10.1007/s12206-019-0723-1

[29] A. Kramschuster, R. Cavitt, D. Ermer, Z. B. Chen, and L. S. Turng, "Effect of processing conditions on shrinkage and warpage and morphology of injection moulded parts using microcellular injection moulding," *Plast. Rubber Compos.*, vol. 35, no. 5, pp. 198–209, 2006. doi: 10.1179/174328906X128199

[30] S. Li et al., "Optimization of injection molding process of transparent complex multi-cavity parts based on kriging model and various optimization techniques," *Arab. J. Sci. Eng.*, vol. 46, no. 12, pp. 11835–11845, Dec. 2021. doi: 10.1007/s13369-021-05724-2

[31] G. J. Kang, C. H. Park, and D. H. Choi, "Metamodel-based design optimization of injection molding process variables and gates of an automotive glove box for enhancing its quality," *J. Mech. Sci. Technol.*, vol. 30, no. 4, pp. 1723–1732, Apr. 2016. doi: 10.1007/s12206-016-0328-x

[32] B. Suksawat, "Development of in-process surface roughness evaluation system for cast nylon 6 turning operation," *Procedia Eng.*, vol. 15, pp. 4841–4846, 2011. doi: 10.1016/j.proeng.2011.08.903

[33] T. J. Ross, *Fuzzy logic with engineering applications*, 3rd ed. West Sussex: John Wiley and Sons Ltd, 2010. doi: 10.1002/9781119994374

[34] M. Nikolić, M. Šelmić, D. Macura, and J. Ćalić, "Bee colony optimization meta-heuristic for fuzzy membership functions tuning," *Expert Syst. Appl.*, vol. 158, pp. 1–10, 2020. doi: 10.1016/j.eswa.2020.113601

[35] P. Tremante, K. Yen, and E. Brea, "Tuning of the membership functions of a fuzzy control system using pattern search optimization method," *J. Intell. Fuzzy Syst.*, vol. 37, no. 3, pp. 3763–3776, 2019. doi: 10.3233/JIFS-190003

[36] M. Moayyedian, K. Abhary, and R. Marian, "Optimization of injection molding process based on fuzzy quality evaluation and Taguchi experimental design," *CIRP J. Manuf. Sci. Technol.*, vol. 21, pp. 150–160, May 2018. doi: 10.1016/J.CIRPJ.2017.12.001

[37] M. L. Chaves, L. Sánchez-González, E. Díez, H. Pérez, and A. Vizán, "Experimental assessment of quality in injection parts using a fuzzy system with adaptive membership functions," *Neurocomputing*, vol. 391, pp. 334–344, May 2020. doi: 10.1016/j.neucom.2019.06.108

[38] K. T. Chiang and F. P. Chang, "Application of grey-fuzzy logic on the optimal process design of an injection-molded part with a thin shell feature," *Int. Commun. Heat Mass Transf.*, vol. 33, no. 1, pp. 94–101, 2006. doi: 10.1016/j.icheatmasstransfer.2005.08.006

[39] Y. Hu and K. Wu, "Application of expert adjustable fuzzy control algorithm in temperature control system of injection machines," *Comput. Intell. Neurosci.*, vol. 2022, 2022. doi: 10.1155/2022/3616814

[40] K. Li, S. Yan, Y. Zhong, W. Pan, and G. Zhao, "Multi-objective optimization of the fiber-reinforced composite injection molding process using Taguchi method, RSM, and NSGA-II," *Simul. Model. Pract. Theory*, vol. 91, pp. 69–82, 2019. doi: 10.1016/j.simpat.2018.09.003

[41] J. Chen, Y. Cui, Y. Liu, and J. Cui, "Design and parametric optimization of the injection molding process using statistical analysis and numerical simulation," *processes*, vol. 11, no. 414, pp. 1–17, 2023. doi: 10.3390/pr11020414

[42] S. M. S. Mukras, H. M. Omar, and F. A. Al-Mufadi, "Experimental-based multi-objective optimization of injection molding process parameters," *Arab. J. Sci. Eng.*, vol. 44, no. 9, pp. 7653–7665, 2019. doi: 10.1007/s13369-019-03855-1

[43] G. Lanzaro and M. Andrade, "A fuzzy expert system for setting Brazilian highway speed limits," *Int. J. Transp. Sci. Technol.*, vol. 12, no. 2, pp. 505–524, Jun. 2023. doi: 10.1016/J.IJTST.2022.05.003

[44] M. Kuntoğlu and H. Sağlam, "ANOVA and fuzzy rule based evaluation and estimation of flank wear, temperature and acoustic emission in turning," *CIRP J. Manuf. Sci. Technol.*, vol. 35, pp. 589–603, 2021. doi: 10.1016/j.cirpj.2021.07.011

[45] V. Sharma, P. Kumar, and J. P. Misra, "Cutting force predictive modelling of hard turning operation using fuzzy logic," *Mater. Today Proc.*, vol. 26, pp. 740–744, 2019. doi: 10.1016/j.matpr.2020.01.018

[46] R. Ramanujam, K. Venkatesan, V. Saxena, R. Pandey, T. Harsha, and G. Kumar, "Optimization of machining parameters using fuzzy based principal component

analysis during dry turning operation of inconel 625 - A hybrid approach," *Procedia Eng.*, vol. 97, pp. 668–676, 2014. doi: 10.1016/j.proeng.2014.12.296

[47] B. Suksawat, "Diameter error prediction using fuzzy logic for cast nylon 6 turning operation," *IERI Procedia*, vol. 10, pp. 76–84, 2014. doi: 10.1016/j.ieri.2014.09.094

[48] J. C. Chen, G. Guo, and W. N. Wang, "Artificial neural network-based online defect detection system with in-mold temperature and pressure sensors for high precision injection molding," *Int J Adv Manuf Tech*, vol. 110, no. 7–8, pp. 2023–2033, Sep. 2020. doi: 10.1007/s00170-020-06011-4

5

State-of-the Art of Artificial Intelligence and Prospectives in Modelling of Plastic Injection Molding

5.1 Introduction

Plastic injection molding process is one of the major ever-evolving manufacturing processes [1]. It has become a key component of material processing and has revolutionized the sector with its precision, efficiency, and versatility. As we gear towards the continued application of technology and innovation to plastic injection molding process, the continuous integration of computer aided engineering (CAE), statistical modelling, and intelligent predictive modelling purpose to shape the future of plastic processing through product quality and process efficiency enhancements which aim at tackling the numerous challenges experienced in the manufacturing process.

In plastic injection molding context, CAE serves as a virtual testing ground which allows manufacturers and scholars to simulate and analyze the manufacturing process. Over the years, the evolving capability of CAE has emphasized its role in accelerating product development through process optimization. Statistical modelling on the other hand is a powerful tool that has been playing a crucial role in systematic analysis and optimization of the injection molding process. Through statistical models, manufacturers have been able to gain insights about the complex interaction among process parameters and their influence on product quality. The analysis of historical data and further identification of process patterns has necessitated informed decision making and optimization of parameters for enhanced product quality. Furthermore, the implementation of artificial intelligence to defect predictive modelling ushered in a new era of technological advancement in plastic injection molding. Machine learning algorithms have enhanced the potential to detect and control various defects prior to occurrence. However, all these modelling techniques are merely an approximate representation of the physical injection molding process hence subject to various prediction errors. For this reason, technological advancements have constantly led to

DOI: 10.1201/9781003492498-5

continuous improvement of these modelling techniques for enhanced predictive capability.

5.2 Current Advances in the Application of AI in Injection Molding and Industry 4.0

The automation of the production industry under Industry 4.0 has been ongoing for a while now. This has attracted research on the use of AI in the different stages and applications of injection molding and the adoption of AI methods and techniques in optimising the production of elements through this method. One of the main challenges facing injection molding is the uncertainty of the process due to the many parameters involved. This leaves quality control, defect detection, and process optimization a challenge to the production of injection molded parts. Artificial intelligence models such as reinforcement learning, and genetic algorithms have been used to enhance quality control and optimise injection molding processes.

Machine learning algorithms are currently being applied to monitor and control the molding variables such as temperature, pressure, and cycle times which has been helping to prevent defects such as air bubbles, warping, or surface imperfections hence improving the product consistency. For instance, in the application of machine learning models in closed loop controls to analyse and evaluate the quality of molded elements, it has been observed that the molding parameters such as temperature can be optimized for higher surface qualities of elements [2]. The study conducted also revealed that AI models can be effectively applied in the control of dimensional accuracy of injection molded parts, and by correlation, the control of their resultant weight.

Additionally, [3] investigated the ability of different machine learning algorithms (k-nearest neighbour, naive Bayes, binary decision tree, and linear discriminant analysis) to effectively predict the quality of multi-cavity injection molding. The study concluded that the different machine language algorithms could be adopted in quality prediction even with little training data, with the decision tree and linear discriminant analysis proving superiority (more than 90 %). Other studies evaluating the application of AI in quality prediction and assessment of injection molded elements have been presented in literature [4–7].

Secondly, AI models have been employed in the prediction of maintenance routines and strategies for injection molding machinery. These models, coupled with performance analysis, can analyse sensor data and historical performance of the machinery to create predictive models for maintenance and forecast equipment failures. For example, [8] employed a data-driven predictive maintenance model to predict the performance of the cooling system of an injection molding machine using data collected from edge computing. The models developed were able to monitor the occurrence of any colling issues which prompted a proactive action. The authors noted that to increase the accuracy of predicted performance of elements, accumulation of more training data is crucial. A similar study was conducted [9] on injection molding machines to predict the occurrence of faults and anomalies in their performance through cognitive analytics and learned systems. This study revealed that with big data in industry 4.0, it is possible to create models that receive real-time data and carry out anomaly detection in the elements of injection molding machines. These cognitive models can be used to monitor the performance of injection molding machines and retrain the models to increase the accuracy of prediction [9].

Thirdly, AI has found applications in the design of injection molds and simulation of the injection molding process through CAE. Generative design algorithms can be used to optimise mold designs for specific production requirements. Simulation of mold filling simulations can predict potential defects and suggest modifications. These simulation models, coupled with digital twins, which allow for the creation of virtual replicas of the injection molding process, have assisted molding experts to simulate, test, and optimise the injection molding process without conducting actual experimentation [10]. Lastly, AI models have been employed to select the best materials and parameters for specific injection molding applications by evaluating the material properties, cost, and molding requirements. These models also allow the production experts to evaluate the product requirements and link them to the production process using data.

5.3 Prospects of CAE Modelling

Amidst a wide range of advantages associated with the application of computer aided engineering modelling to plastic injection molding, there are limitations associated with these models that researchers and CAE package developers have been constantly trying to address. These limitations result majorly from the complex nature of the plastic injection molding process in terms of the complex polymer material properties and mold design, and the difficulties associated with modelling such a process [11]. The injection molding process is characterized by heat transfer mechanisms, transient processes with dynamic phase changes and time-varying boundary conditions which adds a layer of complexity to the modelling process. Moreover, CAE models must also take into consideration the complex interaction between polymer material properties and the part geometry. These factors greatly contribute to the challenges associated with accurate modelling and prediction of the injection molding process behavior and demands a comprehensive understanding of the relationships between thermal dynamics, material characteristics and geometric intricacies.

To this effect, CAE modelling packages have undergone constant significant advancements with an aim of improving the reliability and accuracy of the modelling results. There has been continuous adoption of more sophisticated numerical methods and algorithms capable of capturing the complex relationship among injection molding process variables. The models are continuously striding towards a more realistic representation of the injection molding process as a result of the incorporation of dynamic models that are capable of accounting for phase changes and time-varying boundary conditions. For instance, Moldex3D have incorporated a transient cooling analysis feature that takes into consideration the dynamic cooling environment of the injection molding process [12]. Continued research efforts are expected to lead to even more sophisticated numerical methods and algorithms aimed at refining the accuracy of the results.

The underlying assumptions and model simplifications associated with these models in first principle could sometimes result to larger variations between numerically obtained results and actual results [13]. Earlier mathematical models developed involved simpler geometries and were limited to one-dimensional cases coupled with numerous assumptions and model simplifications. However, with the advancement in computer technology and the increasing need for geometrically complex products, three-dimensional numerical models have been introduced. This advancement in solver capability have also reduced the various assumptions and simplifications. Continued research is expected to give rise to more complex mathematical models with limited assumptions for a realistic representation of the physical system.

Also, the eminent effect of mesh element features to the simulation results portrays a major concern to the accuracy of the CAE simulation results. Variations in mesh element size, density and distribution have a significant impact to the simulation results. In CAE modelling, it is important to keep the number of mesh elements maximum and mesh element size minimum for accurate results [14]. However, the choice on the number and size of mesh elements is limited to the computational capability of the solver, the necessary computational time and computer memory and processor capabilities. This has necessitated the need to find a trade-off between the mesh discretization and computational efficiency through a mesh convergence test. This test has become a fundamental requirement for reliable CAE simulations. This test is merely a compromise and may not entirely solve the problem at hand as achieving a true convergence remains elusive. Therefore, mesh-free CAE technology offers a remarkable potential solution to mesh associated challenges [15]. Hence in future, the adoption of this technology by CAE modelling packages would contribute to accurate process modelling and prediction [15].

The application of artificial intelligence and machine learning may play a pivotal role in enhancing the functionalities of the CAE models through optimization of the process based on real-time data and process feedback. Advancements in parallel and cloud-based computing may contribute to faster simulation timeframes. Also, the integration of experimental data into the CAE toolbox could be a promising avenue for the in-software numerical result validation.

Various CAE modelling software packages have incorporated statistical tools such as design of experiments within the interface to enhance parameter analysis and optimization. With the specification of the target process parameters, parameter levels and target quality indices, an experiment can be automatically generated, and data analyzed from the CAE platform. This is a promising milestone and in future, an addition of artificial intelligence to this integration would be substantial. Currently, the mold design process is iterative and involves part and mold design, simulation and optimization, and the adjustment of the designs with respect to the simulation results. With the integration of CAE, statistics, and artificial intelligence in future, it would be possible for a CAE simulation to self-analyze, optimize, and self-adjust the design features of the part geometry or injection mold to suit the optimization requirements.

5.4 Prospects in Statistical Modelling

Statistical modelling has been instrumental in systematic plastic injection molding process evaluation and optimization. Various complex statistical tools have been adopted over time to cope with the evolving process complexities. However, the adoption of statistical modelling techniques to plastic injection molding faces challenges associated with the complexity of the injection molding process, non-linear and time-varying polymer material behavior, high dimensional data and interactions among process variables. An accurate statistical representation of these features has been a challenging endeavor.

Plastic injection molding process involves complex interplay among many factors such as mold design, part geometry, material properties, process properties and environmental conditions. Each of these factors contribute to the dynamic nature of the process and hence developing a single statistical model that captures the effects of all the factors to various quality indices has been challenging. Additional layer of complexity is added onto the models as a result of the non-linear behavior of the polymer material during the molding process and the presence of a large number of variables and parameters at play.

As plastic injection molding process involves a wide range of quality indices and associated defects, statistical optimization of the process has evolved from single objective to multi-objective. However, the efficacy of the multi-objective optimization greatly relies on the number of defects and optimization objectives. Introduction of many objectives to the problem leads to complexities that reduces the efficacy of the solution. To obtain a holistic evaluation and optimization of the process, various studies have adopted a technique of creating a single composite index which is a function of all the weighted defects [16]. However, defect weighting has been more subjective and, in most cases, based only on the opinion of molding experts. In future, adoption of advanced data driven models to track the implications of a particular defect from product formation to product end of life would be necessary. This would enhance a systematic and objective formulation of defect weightings and the implications of the particular defect throughout the product service life [17].

Therefore, addressing issues related to the modelling of complex manufacturing processes such as plastic injection molding remains a focal point for researchers and engineers seeking to enhance the effectiveness of statistical modelling to the process.

5.5 Prospects in AI Modelling

Predictive modelling has been one of the greatest milestones as far as application of artificial intelligence to plastic injection molding process is concerned. Through AI, PIM defects have been modelled, predicted and controlled thereby ushering in a new era of precision and efficiency in manufacturing [2]. AI algorithms have proven suitable to capturing the complex nature of plastic injection molding process. This has been a game-changer and has enabled a thorough understanding of the relationships between the process factors and defects.

With advancements in computer technology, more advanced and complex algorithms capable of accurately modelling the complex nature of manufacturing processes such as injection molding process are anticipated [18]. The ongoing development of AI-based predictive modeling strategies signifies a continuous evolution towards a more robust and adaptable solutions capable of handling the intricacies and complexities of modern manufacturing. Furthermore, the adoption of AI to other dimensions of PIM process such as CAE modelling, statistical modelling, real-time monitoring and adaptive control are anticipated. The integration of AI into these areas holds the promise of comprehensive optimization that would allow manufacturers to not only predict and prevent defects, but also to fine-tune and enhance various aspects of the overall production cycle.

5.6 Conclusion

Therefore, the continuous evolution of plastic injection molding process has been significantly driven by advancements in CAE modelling, statistical modelling, and AI-based predictive modelling. Furthermore, the integration of AI into CAE and statistical modelling have a potential of providing a leverage for enhanced process evaluation and optimization.

In the realm of CAE modeling, AI would contribute to a more accurate virtual experimentation and design optimization thereby enabling manufacturers to accurately explore a wide range of design and process configurations. Integration of statistical modelling with AI would offer a more powerful tool for understanding the relationships among different variables and result to more improved process control and defect prevention initiatives. Real-time monitoring and adaptive control powered by AI would facilitate dynamic adjustments during the injection molding process and ensure rapid responses to changing conditions thereby minimizing the risk of defects.

As the trajectories of CAE, statistical and AI modelling of plastic injection molding process continues to rise, the industry can anticipate holistic transformations in defect control and process optimization. This is poised to usher in an era of unprecedented precision, efficiency and adaptability in plastic injection molding process.

References

[1] M. Czepiel, M. Bańkosz, and A. Sobczak-Kupiec, "Advanced injection molding methods: Review," *Materials*, vol. 16, no. 17, p. 5802, Aug. 2023. doi: 10.3390/ma16175802

[2] S. S. Aminabadi et al., "Industry 4.0 in-line AI quality control of plastic injection molded parts," *Polymers (Basel).*, vol. 14, no. 3551, Sep. 2022. doi: 10.3390/polym14173551

[3] R. D. Parizs, D. Torok, T. Ageyeva, and J. G. Kovacs, "Machine learning in injection molding: An industry 4.0 method of quality prediction," *Sensors*, vol. 22, no. 2704, 2022. doi: doi.org/10.3390/s22072704

[4] M. Gülçür and B. Whiteside, "A study of micromanufacturing process finger-prints in micro-injection moulding for machine learning and Industry 4.0 appli-cations," *Int. J. Adv. Manuf. Technol.*, vol. 115, no. 5–6, pp. 1943–1954, 2021. doi: 10.1007/s00170-021-07252-7

[5] J. Y. Chen, C. C. Tseng, and M. S. Huang, "Quality indexes design for online monitoring polymer injection molding," *Adv. Polym. Technol.*, vol. 2019, 2019. doi: 10.1155/2019/3720127

[6] K. C. Ke and M. S. Huang, "Quality classification of injection-molded com-ponents by using quality indices, grading, and machine learning," *Polymers (Basel).*, vol. 13, no. 3, pp. 1–18, 2021. doi: 10.3390/polym13030353

[7] H. S. Park, D. X. Phuong, and S. Kumar, "AI based injection molding process for consistent product quality," *Procedia Manuf.*, vol. 28, pp. 102–106, 2019. doi: 10.1016/j.promfg.2018.12.017

[8] S. Farahani, V. Khade, S. Basu, and S. Pilla, "A data-driven predictive mainte-nance framework for injection molding process," *J. Manuf. Process.*, vol. 80, no. June, pp. 887–897, 2022. doi: 10.1016/j.jmapro.2022.06.013

[9] V. Rousopoulou, A. Nizamis, T. Vafeiadis, D. Ioannidis, and D. Tzovaras, "Predictive maintenance for injection molding machines enabled by cognitive analytics for industry 4.0," *Front. Artif. Intell.*, vol. 3, no. November, pp. 1–12, 2020. doi: 10.3389/frai.2020.578152

[10] M. R. Khosravani, S. Nasiri, and T. Reinicke, "Intelligent knowledge-based sys-tem to improve injection molding process," *J. Ind. Inf. Integr.*, vol. 25, no. August 2021, p. 100275, 2022. doi: 10.1016/j.jii.2021.100275

[11] P. Kennedy and R. Zheng, *Flow analysis of injection molds*, 2nd ed. Munich: Hanser Publications, 2013. doi: 10.3139/9781569905227

[12] A. Torres-Alba, J. M. Mercado-Colmenero, J. D. D. Caballero-Garcia, and C. Martin-Doñate, "A hybrid cooling model based on the use of newly designed fluted conformal cooling channels and fastcool inserts for green molds," *Polymers (Basel).*, vol. 13, no. 18, p. 3115, Sep. 2021. doi: 10.3390/polym13183115

[13] H. Zhou, *Computer modeling for injection molding: simulation, optimization, and control*, 1st ed. New Jersey: John Wiley and Sons Inc, 2013. doi: 10.1002/9781 118444887

[14] D. A. De Miranda, "Influence of mesh geometry and mesh refinement on math-ematical models of thermoplastic injection simulation tools," *IOSR J. Mech. Civ. Eng.*, vol. 15, no. 3, pp. 38–44, 2018. doi: 10.9790/1684-1503013844

[15] L. Veltmaat, F. Mehrens, H. J. Endres, J. Kuhnert, and P. Suchde, "Mesh-free simulations of injection molding processes," *Phys. Fluids*, vol. 34, no. 3, 2022. doi: 10.1063/5.0085049

[16] M. Moayyedian, K. Abhary, and R. Marian, "Optimization of injection molding process based on fuzzy quality evaluation and Taguchi experimental design," *CIRP J. Manuf. Sci. Technol.*, vol. 21, pp. 150–160, May 2018. doi: 10.1016/J.CIRPJ. 2017.12.001

[17] C. Li, Y. Chen, and Y. Shang, "A review of industrial big data for decision mak-ing in intelligent manufacturing," *Eng. Sci. Technol. Int J.*, vol. 29, p. 101021, May 2022. doi: 10.1016/j.jestch.2021.06.001

[18] J. F. Arinez, Q. Chang, R. X. Gao, C. Xu, and J. Zhang, "Artificial intelligence in advanced manufacturing: Current status and future outlook," *J. Manuf. Sci. Eng. Trans. ASME*, vol. 142, no. 11, Nov. 2020. doi: 10.1115/1.4047855

Index

Pages in *italics* refer to figures and pages in **bold** refer to tables.

fused deposition modelling, 1
fuzzification, 76, 81
fuzzy inference system, 76–98
fuzzy logic, 75–99
fuzzy sets, 76

G

gating suitability test, 15
gaussian membership function, 78–95

H

HDPE, 14
Hele-Shaw model, 18–19

I

imprecision, 75
industry 4.0, 105–106
inertial force, 23
injection molding, 1–8
injection pressure, 37–40, 48–**67**, 78–93
injection unit, 2, 3
intelligent approximation algorithms, 72
interaction effects, 53–**67**
internal stresses, 48
intuition, 76

K

Kriging model, 73–75

L

laminar flow, 23
linguistic rules, 76–95

M

machine control, 3
machine learning, 7, 104
mains effects, 48–**67**
Mamdani fuzzy inference system, 78
material backflow, 53
material manufacturing, 1

MATLAB fuzzy logic designer toolbox, 77
melt temperature, 37–40, 48–**67**, 78–93
membership function, 76–98
membership function plot, 83–87
mesh convergence test, 25
mesh element, 107
mesh refinement, 25
model overfitting, 55–66
modified Tait equation, 20
mold cavity, 1
mold design validation, 33
moldex3D, 12, 77
molding defect index, 65
mold simulation, 8
mold temperature, 37–40, 48–**67**
molecular arrangement, 48
molecular flow orientation, 6
molecular structure, **7**
multivariability, 72

N

net plastic manufacturing process, 2
Newtonian model, 33
non-linearity, 72
numerical data validation, 37
numerical density, 29–32
numerical simulation, 73

O

optimization, 44, 76
optimize, 11

P

packaging bottle cap, 12
packing, 2–7
packing phase, 20
packing pressure, 37–40, 48–**67**, 78–93
packing time, 37–40, 48–**67**, 78–93
parameter causality, 86
parameter screening, 76–78
Pareto chart, 48–66
pattern search, 76